MECHANICAL VIBRATION METHODS FOR STUDYING PHYSICAL PROPERTIES OF SOLID MATERIALS

MATERIALS SCIENCE
AND TECHNOLOGIES

Additional books in this series can be found on Nova's website under the Series tab.

Additional E-books in this series can be found on Nova's website under the E-books tab.

MATERIALS SCIENCE AND TECHNOLOGIES

MECHANICAL VIBRATION METHODS FOR STUDYING PHYSICAL PROPERTIES OF SOLID MATERIALS

Y. HIKI

Novinka
Nova Science Publishers, Inc.
New York

For permission to use material from this book please contact us:
Telephone 631-231-7269; Fax 631-231-8175
Web Site: http://www.novapublishers.com

NOTICE TO THE READER

The Publisher has taken reasonable care in the preparation of this book, but makes no expressed or implied warranty of any kind and assumes no responsibility for any errors or omissions. No liability is assumed for incidental or consequential damages in connection with or arising out of information contained in this book. The Publisher shall not be liable for any special, consequential, or exemplary damages resulting, in whole or in part, from the readers' use of, or reliance upon, this material.

Independent verification should be sought for any data, advice or recommendations contained in this book. In addition, no responsibility is assumed by the publisher for any injury and/or damage to persons or property arising from any methods, products, instructions, ideas or otherwise contained in this publication.

This publication is designed to provide accurate and authoritative information with regard to the subject matter covered herein. It is sold with the clear understanding that the Publisher is not engaged in rendering legal or any other professional services. If legal or any other expert assistance is required, the services of a competent person should be sought. FROM A DECLARATION OF PARTICIPANTS JOINTLY ADOPTED BY A COMMITTEE OF THE AMERICAN BAR ASSOCIATION AND A COMMITTEE OF PUBLISHERS.

LIBRARY OF CONGRESS CATALOGING-IN-PUBLICATION DATA

Available upon Request
ISBN: 978-1-61728-286-7

Published by Nova Science Publishers, Inc. ✦ *New York*

CONTENTS

PREFACE

The method of observing the response of a solid to an applied mechanical vibration - mechanical vibration method - is useful for investigating various physical properties of materials. The present article is intended to show the utility of the method. First, the formalism of the elasticity and mechanical vibration is concisely described. Second, our previous research mainly concerning crystalline solids is presented. Topics are such as characterization of defects in crystals, nonlinear elasticity of crystals, electron conduction and superconductivity, and melting phenomenon. Internal friction and ultrasound propagation experiments are used. Third, our recent research concerning amorphous or glassy materials is presented. The materials are inorganic, organic, and metallic glasses of various kinds. The phenomena of the glass transition and crystallization are the main objects. Wide range of measurement frequencies is adopted in the experiments. Viscosity, internal friction, ultrasound propagation, and Brillouin scattering experiments are used.

Chapter 1

INTRODUCTION

Contents of the present article are as follows. In chapter II, basic concepts and formalism of the elasticity and elastic vibration are given. The description is limited to the items which are useful for understanding the following main body of the article, since details of these items can be referred to other standard text books. In chapter III, our previous research (before 1990) is presented. The materials used are various kinds of crystalline solids. Topics presented are divided into three categories: characteristics of imperfections in defect crystals, lattice anharmonicity or elastic nonlinearity of crystals, and other topics such as electron conduction, superconduction, and melting transition. The presentation of these is rather brief and simple. In chapter IV, our recent research (after 1990) is presented. The materials used are several kinds of amorphous or glassy solids. The object of these studies is mainly aimed to the phenomena of glass transition and crystallization. Especially, the glass transition is interesting for us since there are still many unsolved problems concerned. A wide range of measurement frequency is adopted in these mechanical vibration studies, which is a characteristic feature of our study. The presentation of these is rather more precise since we are presently continuing these studies.

Chapter 2

PRELIMINARY

ABSTRACT

In a linear elastic medium, the stress and the strain are defined in the tensor form. The Hooke's law representing the linear elasticity is given in the tensor form, and further in the matrix form, to define the elastic stiffness constants. The elastic constants are also defined as the second-order strain derivatives of thermodynamic functions. Higher-order elastic constants are defined as the higher-order strain derivatives (A). The mechanical damping loss occurring in a vibrating body and the attenuation of waves propagating in the body are taken into consideration. Various quantities representing the damping loss, or the internal friction, and quantities representing the attenuation of waves are given. The relations between these quantities are shown (B).

A. ELASTICITY

The linear elasticity of a crystal is firstly described. Three-dimensional Cartesian coordinates x_i ($i = 1, 2, 3$) are taken in a body. Let the displacement components of a material point be u_i. The strain ε_{ij} of the body at the point is defined as

$$\varepsilon_{ij} = (1/2) \left[(\partial u_i/\partial x_j + \partial u_j/\partial x_i) \right]. \tag{1}$$

The stress τ_{ij} acting on the material point is defined as

$\tau_{ij} = j$-th component of force per unit area
acting on the plane perpendicular to i-axis.   (2)

The strain and the stress are shown to be the symmetric second-rank tensors. The Hooke's law showing the linear elasticity is represented as

$$\tau_{ij} = c_{ijkl}\varepsilon_{kl} \ (i, j, k, l = 1, 2, 3), \tag{3}$$

where c_{ijkl} is a fourth-rank tensor and is called the elastic stiffness. In such a tensor representation, when repetition of index occurs in a term a summation is always taken for the index. Further, from the symmetry of these tensors we can use the contraction of their indexes:

$$(ij) = (11) \rightarrow I = 1 \text{ etc}, (ij) = (23) = (32) \rightarrow I = 4 \text{ etc}. \tag{4}$$

Then the alternative form of the Hooke's law is

$$(\tau_I) = (c_{IJ})(\varepsilon_J) \ (I, J = 1 - 6), \tag{5}$$

where (τ_I) and (ε_J) are 1×6 matrixes, and (c_{IJ}) is a 6×6 matrix. The number of independent and non-zero components of the stiffness is 21. Further, the crystal symmetry reduces the number. For example, there are three independent non-zero components for the cubic system, c_{11}, c_{12}, and c_{44}.

Furthermore, from a thermo-mechanical argument, the elastic stiffness can be defined as the second-order strain derivative of the thermodynamic functions such as the internal energy and the Helmholtz free energy. Therefore, the usual elastic stiffness constant is called the second-order elastic constant. The higher-order elastic constants are defined in the same manner. For example, for the cubic system the numbers of the third-order (C_{IJK}) and the fourth-order (C_{IJKL}) elastic constants are 6 and 11, respectively.

B. MECHANICAL VIBRATION

Consider a mechanical vibration applied to a solid. Depending on the shape of the solid and the mode of the vibration, intensive resonance vibrations occur at definite vibration frequencies. From such an observation the elastic constant or their combination can be determined. For example, from the longitudinal vibration the Young's modulus which is a combination of several elastic constants is obtained. From the torsion vibration the shear elastic constant is obtained. These experimental methods are familiar and further description may not be necessary.

Now the phenomenon of the mechanical damping loss in a material is taken into account. Consider a body in a state of forced vibration in a certain vibration mode. Let the representative displacement (elongation, torsion, etc) be θ, the inertial quantity (mass, moment of inertia, etc) I, the force externally applied F_a, and the force produced in the body F_p. The equation of motion of the body is

$$I(d^2/dt^2)\theta = F_a + F_p; \quad F_a = F_0 e^{i\omega t}, \quad F_p = - k^*\theta, \tag{6}$$

where ω is the angular frequency of vibration. When a mechanical loss exist, F_p is retarded from F_a, and k^* is a complex elastic modulus, which is written as

$$k^* = k(1 + i\varphi), \tag{7}$$

where φ is the phase-difference angle, which is sometimes denoted by δ. Assume the solution of the equation of motion as $\theta = \theta_0^* e^{i\omega t}$, then we obtain the absolute value of the complex amplitude θ_0^* as a function of the frequency ω. A peak appears at $\omega_0 = (k/I)^{1/2}$ and the half width of the peak is

$$\text{half width} = 3^{1/2}\varphi. \tag{8}$$

Next, a vibrating body decaying freely is considered. The solution of the equation of motion is obtained by setting $F_a = 0$ in Eq. (6) and assuming the solution $\theta = \theta_0^* e^{i\omega t}$. Then the logarithmic decrement defined as the logarithm of the ratio of the neighboring amplitudes of vibration is obtained as

$$\text{log.dec.} = \pi\varphi. \tag{9}$$

Now the internal friction (IF) is defined as

$$IF = \Delta E/2E, \tag{10}$$

where E is the time-averaged vibration energy and ΔE is the energy dissipated in a cycle. After considering that the vibration energy is proportional to squared amplitude of vibration, we obtain the relation

$$IF = \pi\ \varphi. \tag{11}$$

Sometimes the inverse of the quality factor Q^{-1} define as

$$Q^{-1} = \varphi \tag{12}$$

is used to represent the internal friction.

Another type of IF can also be considered. When the stress applied to a solid is increased and decreased and a static hysteresis exists in the stress-strain relation, a mechanical loss proportional to the area of the hysteresis curve occurs. The loss also occurs in the case of vibration stress. Such a kind of IF is called the hysteresis-type loss, which is strongly dependent on the strain amplitude of vibration but is independent of the vibration frequency.

Meanwhile, observation of the attenuation of propagating elastic waves in a body is also useful to obtain the damping characteristic of the material. As the waves propagate distance x its amplitude is decreased by a factor $e^{-\alpha x}$, where α is the attenuation coefficient and the unit is usually represented by nepers/cm. Sometimes the decibel (dB) unit is also used:

$$1\ \text{nepers/cm} = 8.686\ \text{dB/cm}. \tag{13}$$

It is shown that the following relation holds between IF and the attenuation:

$$IF = \alpha\lambda, \tag{14}$$

where λ is the wavelength of the sound.

Chapter 3

PREVIOUS RESEARCH

ABSTRACT

Amplitude dependent kHz internal friction was measured for Pb (A) and quartz (B) crystals, and the results were explained by the mechanism of hysteresis-type IF due to pinned dislocations. Frequency dependent MHz sound attenuation was measured for LiF, NaCl, and Al crystals under hydrostatic pressure. The results were explained by the mechanism of resonance-type IF due to dislocations, and the dislocation-phonon interaction was discussed (C). Sound attenuation in solid ^{4}He was studied to find the quantum nature of motions of dislocations and point defects (D). Melting of solid ^{4}He (E) and H_2O ice (F) was studied by sound attenuation measurement, and a rapid increase of dislocation density near melting point was found. Sound attenuation in high-purity normal-conducting Al was measured, and mechanical losses due to normal electrons and interaction of electron and dislocation were studied (G).

Attenuation measurement was carried out in superconducting Al to determine the anisotropic energy gap in the superconductor (H). By measuring the changes of sound velocity under uniaxial and hydrostatic pressure the third-order elastic constants were determined for Cu, Ag, and Au (I). Experiment of production of a sound by non-collinear collisions of two sounds was performed to study three-phonon process due to lattice anharmonicity (J). Nonlinear stress-strain relation was studied in Cu whiskers having high yield stress and wide elastic range, and the results were compared with the conclusion of higher-order elasticity theory (K).

A. INTERNAL FRICTION OF LEAD

IF of Pb single crystals (99.99% pure) was measured with longitudinal vibration in kHz range [1, 2]. The apparatus used is a composite piezo-electric oscillator [3], by which the values of the IF and the strain amplitude of vibration can be measured at several vibration frequencies at various temperatures.

IF value δ is represented as a sum of the amplitude-independent part and the amplitude-dependent part, $\delta = \delta_i + \delta_h$. These two parts can be written as [4]

$$\delta_i \propto \Lambda L_c^4 B\omega,$$
$$\delta_h = (c_0/\varepsilon^{1/2}) \exp(-c_1/\varepsilon);$$
$$c_0 \propto \Lambda/L_c^{3/2}, c_1 \propto 1/L_c, \tag{15}$$

where Λ is the dislocation density, L_c is the distance between the pinning points or the pinning length, B is a quantity called the damping constant for dislocation motion, ω is the angular frequency of vibration, and ε is the strain amplitude.

The results and analysis of the amplitude dependence of IF are here presented. The origin of such IF is a hysteresis-type damping loss due to the depinning and repining of crystal dislocations by impurity pinning points. IF monotonously increases with increasing strain amplitude, and is independent of vibration frequency since the loss is the hysteresis type. Such frequency independence can be seen from the experimental result shown in figure 1. Meanwhile, Eq. (15) shows that $\log(\varepsilon^{1/2}\delta_h)$ is proportional to ε^{-1}. This relation can be seen in the experimental result shown in figure 2 (Granato-Lücke plot).

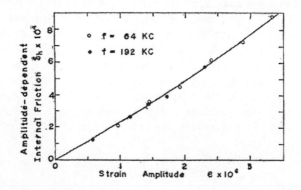

Figure 1. Amplitude dependent internal friction measured at two frequencies.

As an example of the utility of the study of amplitude dependent IF, the diffusion of impurity atoms is investigated. Under the influence of the stress field of a dislocation, impurity atoms migrate to the dislocation line. The number of impurity atoms arrive at the unit length of dislocation line within time t is [5]

$$n(t) = a_1(Dt/RT)^{2/3},\tag{16}$$

where a_1 is a constant, $D = D_0 \exp(-Q/RT)$ is the diffusion constant, Q is the activation energy of diffusion, and R and T have their usual meanings. After considering that $n(t) = [L_c(t)]^{-1} - [L_c(0)]^{-1}$, we obtain the relation

$$c_1(t) - c_1(0) = a_2[\exp(-Q/RT)/T]^{2/3}t^{2/3},\tag{17}$$

where a_2 is a constant. The results of the plots using Eq. (17) and the data in figure 2 are shown by the line a in figure 3. The data for other annealing temperatures are also shown by the lines b, c, and d. Using these results the value of the activation energy Q is determined as shown in figrue 4. A variety of other aspects of IF in Pb have also been studied and these can be seen in the original articles [1, 2]. It is finally noted that the IF value of Pb due to the dislocation damping is rather large and the IF study can conveniently be carried out in this material.

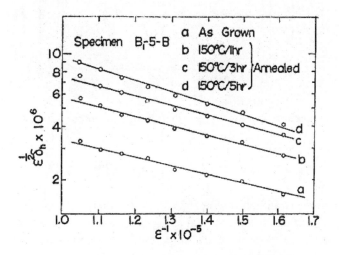

Figure 2. Analysis of amplitude dependence of internal friction.

Y. Hiki

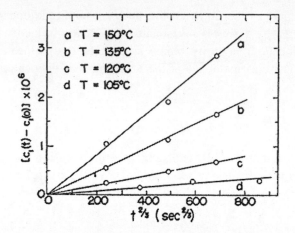

Figure 3. Analysis of effect of annealing upon internal friction.

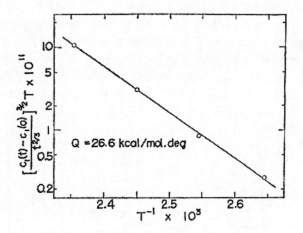

Figure 4. Determination of activation energy for diffusion of point defect.

B. INTERNAL FRICTION OF QUARTZ

IF of natural Brazilian quartz [6] and hydro-thermally synthesized quartz [7] has been studied. Quartz crystal is used as the piezo-electric transducer, and such studied are practically important. A number of specimens of cylindrical form having various crystallographic orientations are used for the composite piezo-electric measurement of IF. For the quartz crystal, the orthogonal X(electric)-axis, Y(mechanical)-axis, and Z(optical)-axis are chosen. The axis of the specimen is in the YZ-plane, and let the angle between the specimen axis and the Y-axis be θ. When IF is mainly due to the dislocation damping and the slip plane of dislocation is parallel to the Y-axis, IF value is proportional to the orientation factor

$$s = a_1 \sin\theta \cos\theta, \tag{18}$$

where a_1 is a constant depending on the slip direction of the dislocation.

In figure 5 the Granato-Lücke plot is shown for the case of the natural quartz specimens with different orientations. In each specimen the line slope is different, which shows that the pinning length is different. In figure 6 the values of the amplitude-independent part δ_i are plotted against the angle θ for a number of specimens. As can be expected, the δ_i-vs-θ is almost well represented by Eq. (18). The scatter of the data points can be due to the differences of the pinning length and the dislocation density in specimens.

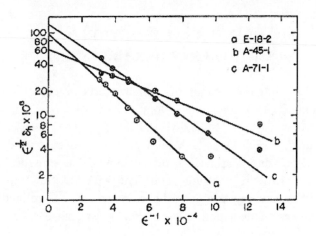

Figure 5. Analysis of amplitude dependence of internal friction.

The dislocation density is an important quantity for qualifying the material for using as the piezo-electric transducer. The lower the dislocation density, the lower the IF value (higher Q value). All IF data obtained have been analyzed precisely [6, 7] and the following result was obtained. The dislocation density is $\Lambda = 10^3 - 10^4$ cm^{-2} in natural quartz specimens and $\Lambda = 10^2 - 10^3$ cm^{-2} in synthetic quartz specimens. The result seems to be practically quite useful.

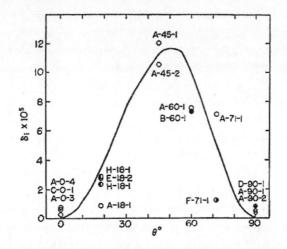

Figure 6. Amplitude independent internal friction for various specimens.

C. INTERNAL FRICTION OF
CRYSTALS UNDER PRESSURE

Pressure is an important parameter which affects almost all of the physical properties of materials. Attenuation of longitudinal MHz pulsed ultrasound has been measured in LiF and NaCl [8] and Al crystals [9] under hydrostatic pressure up to $P = 3500$ kg/cm^2. A high pressure vessel is used, in which a hydrostatic oil pressure is applied to the specimen. The pressure is measured by a calibrated Mn-wire pressure gage. The LiF and NaCl specimens are optical-grade Harshaw crystals, and Al specimens are zone-refined 99.9999 % crystals. The sound attenuation is measured by the pulse reflection method using a Matec apparatus.

IF (Δ) depends remarkably on vibration frequency f. An example of the results is shown in figure 7 for the case of LiF at zero pressure. The result of the theory of dislocation damping [4] is here shortly presented. The amplitude independent IF is taken, since the amplitude of sound used is very small. The equation of motion of the dislocation line between pinning points is considered, where the damping term, $B \times$ velocity, is included and here B is called the damping constant. IF is shown to be represented as

$$\Delta = (a_1 \Lambda L^2)(\omega \tau)/[1 + (\omega \tau)^2]; \quad \tau = a_2 BL^2, \tag{19}$$

where a_1 and a_2 are constants, Λ is the dislocation density, L is the pinning length, B is the damping constant, and ω is the angular frequency. The parameter fitted curve using Eq. (19) is shown in figure 7. There appears a broad peak, which is called the overdamped (high damping) resonance peak. Here, the peak height is $a_1 \Lambda L^2$, and the peak position is given by the relation $\omega \tau = 1$.

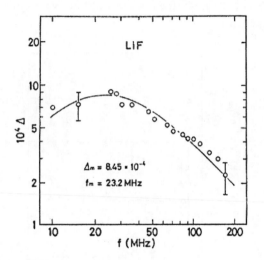

Figure 7. Frequency dependence of internal friction.

Now, IF measurements under pressure are carried out for LiF, NaCl, and Al specimens. In figure 8 the rate of change of IF with pressure P is shown at various vibration frequencies f for LiF and NaCl.. From such data we can calculate the pressure dependence of the damping constant, $(1/B_0)(dB/dP)$. It is considered that the damping constant represents a quantity controlled by the interaction of the moving dislocation and the phonons (quantized lattice vibration) existing in the material. We have theoretically calculated $(1/B_0)(dB/dP)$ value [10] on the basis of various damping mechanisms such as the phonon scattering [11], phonon viscosity [12], fluttering mechanism [13], and so on. The results have been compared with our experimental data. Details can be seen in the article [10].

The same kind of pressure dependence study was also made for the case of Al [9, 10]. The experimental results are shown in figure 9 for deformed specimen (State I) and the specimen irradiated by neutrons in a reactor (State II). The specimens are aged at room temperature, and the aging time is represented by t in the figure. The Al specimens originally contain Si as the

main impurity, and Si atoms are further introduced in the specimen by the
irradiation through a nuclear reaction. By the aging treatment, the dislocation
density decreases, and the pinning length decreases due to the impurity
migration. It is then expected that the peak height decreases and the peak
position shifts to higher frequency (see Eq. (19)). These expectations are truly
realized as can be seen in figure 9. Above effects are more markedly seen in
the State II case, since the impurity content in State II specimen is higher than
that in the State I specimen.

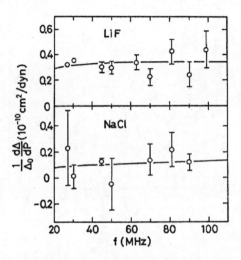

Figure 8. Rate of change of internal friction with pressure as a function of frequency.

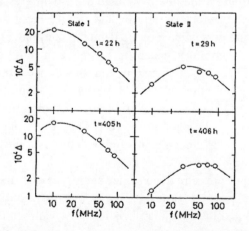

Figure 9. Change of overdamped resonance peak with aging time for specimen in
deformed (State I) and irradiated (State II) specimens.

D. INTERNAL FRICTION OF SOLID HELIUM

Helium does not freeze under its saturated vapor pressure even at very low temperature, and this is a consequence of the quantum-mechanical large-amplitude zero-point motion of atoms. Thus the solid helium is labeled as a quantum solid. The study of IF has been carried out for such an interesting solid [14, 15]. Hexagonal closed packed (hcp) ^{4}He is chosen as the specimen material, and a crystal is grown in a pressurized low-temperature cell. Two kinds of specimens, grown at 32.5 and 60.0 atm and having the molar volume of 20.5 and 19.2 cm^3/mole, respectively, are used. A large number of crystals have been gown and about one fourth of them were good single crystals available for the IF experiment. The sound attenuation measurement is performed by the pulse reflection or the pulse transmission method using the Matec apparatus.

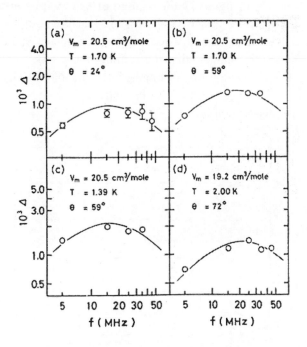

Figure 10. Frequency dependence of internal friction for several specimens.

The results of frequency dependence of IF for several specimens are illustrated in figure 10 , and here the overdamped resonance behavior can be seen. The orientation angle θ of the specimen, which is chosen as in the case of quartz crystal, can be determined from the experimental sound velocity value. In the present helium crystals the same orientation dependence of IF as has been seen in figure 6 for quartz crystals is again obtained [14]. Thus the origin of IF in solid helium crystal is certainly the dislocation damping. Various parameters such as the dislocation density and the pinning length can be determined by analyzing the overdamped-resonance data.

It is shown from Eq. (19) that

$$f_m^{-1/2} \propto L, \tag{20}$$

where f_m is the peak frequency of the overdamped resonance, and L is the pinning length. Now, the following experiment is carried out. The temperature of a specimen is suddenly changed and is kept at another temperature, the overdamped resonance data is successively taken, and the change of the data with time is observed. In figure 11 the change of $f_m^{-1/2}$ with elapsed time t is shown. From such an experiment it can be seen that the equilibrium pinning length L is increased or decreased when temperature is lowered or enhances, respectively. The results suggest that the pinning cannot be due to impurity atom, because the concentration of impurity atoms on dislocation line is proportional to $\exp(E/k_BT)$ where E is the interaction energy between the impurity atom and dislocation line.

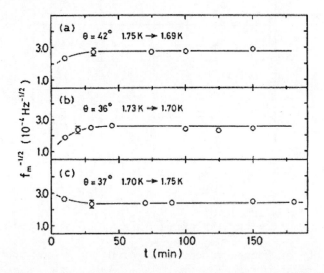

Figure 11. Changes of $f_m^{-1/2}$ with time after temperature is suddenly decreased (a), (b) and increased (c).

F. MELTING OF ICE CRYSTAL

Ice of water is a common solid material but is very special and interesting in its physical properties due mainly to the nature of the bonding between the constituent atoms. The sound attenuation measurement has been performed for ice single crystals near and below the melting temperature [21]. A long fused quartz cell with a rectangular cross section is used to grow the crystal, and a quartz ultrasonic transducer is attached on the outer wall for the ultrasonic measurement. The main impurity in the raw material is 0.40 ppm Na$^+$.

IF of crystals is measured as a function of frequency, and the overdamped resonance peak can be observed. Using the data the dislocation density Λ is determined. The temperature dependence of Λ is shown in figure 16. The dependence is well represented by Eq. (22) as can be seen in the figure. The obtained parameter value is $T_0 = 273.16$ K, which is just the melting temperature as has been expected since ice is a classical, not quantum, solid.

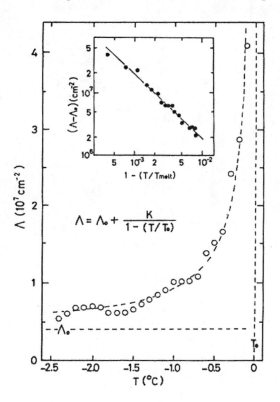

Figure 16. Temperature dependence of dislocation density in ice.

Logarithm of peak frequency f_m of the overdamped resonance is plotted against T^{-1} in figure 17. As can be seen in Eq. (20) f_m is inversely proportional to the square of the pinning length L^2, and L is inversely proportional to the concentration of impurity on dislocation line. The concentration c is given as

$$c = c_0 \exp(E/k_B T), \tag{24}$$

where E is the binding energy between the impurity and dislocation line. Thus the result shown in figure 17 is well understood. Here, solid line represents the result of the least-squares-fit of the data, and data points for various specimens are mostly between two parallel dashed lines. The value of the binding energy obtained is $E = 0.18$ eV. The pinning agencies can be the main impurity Na$^+$.

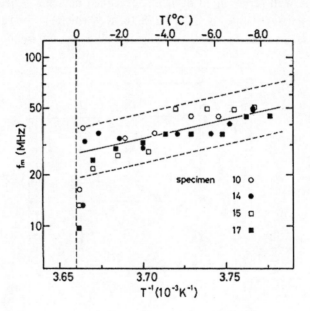

Figure 17. Resonance peak frequency vs inverse temperature.

The effect of impurity doping is further studied. In figure 18 temperature dependence of IF is shown for ice crystals doped with various amounts of Na$^+$ ions. The overall IF values are larger for more heavily doped crystals. IF value in heavily doped crystals gradually and then more rapidly increases with temperature. The results can qualitatively be explained as follows. As previously considered, dislocations are spontaneously produced as the temperature is increased near to the melting point. The dislocations with inverse signs can be annihilated with each other through slipping on the slip plane. The slipping motion is obstructed by the impurities since they act as pinning agencies or as obstacles for the motion. Further discussion concerned, however, may not be so easily made.

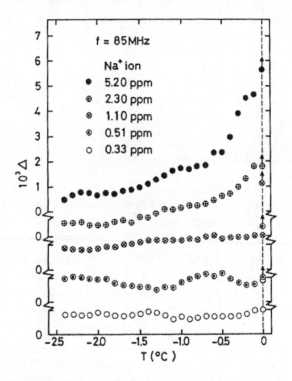

Figure 18. Temperature dependence of internal friction for specimens containing various amounts of Na$^+$ ions.

G. INTERNAL FRICTION IN
NORMAL-CONDUCTING ALUMINUM

IF of high-purity Al (99.9999 %) in the electronic normal-conducting state has been measured by the pulse ultrasonic method [22]. The measurements are performed for annealed and deformed (0.14 – 1.4 %) specimens for studying the effect of dislocations on the low-temperature IF. Figure 19 shows the overall features of the temperature dependence of ultrasonic attenuation α in the annealed and deformed specimens, where solid curves are guide lines. Here, T_c ($\approx$ 1.18 K) is the superconducting transition temperature.

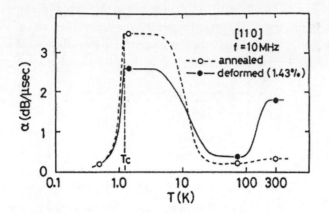

Figure 19. Temperature dependence of ultrasonic attenuation in annealed and deformed crystals.

Characteristics of the sound attenuation are as follows. (a) Room temperature (300 K). The attenuation α is increased when the specimen is deformed. The main origin of IF is the dislocation damping, and the increase of α is due to the increased dislocation density. (b) Liquid nitrogen temperature (77 K). The increment of α in the deformed specimen becomes smaller. The reason is that the dislocation damping constant B (see Eq. (19)) is smaller at lower temperatures, since the damping is due to the dislocation-phonon interaction and the population of phonons decreases with decreasing temperature. (c) Normal-conducting temperature (< 30 K). As the temperature is further lowered, α increases and then becomes constant. Here, the origin of IF is the electron damping. The mechanism of the damping is as follows. The spatial distribution of electrons becomes nonequilibrium when the positive ions in the crystal are displaced by the sound. The electrons relax to a new equilibrium position, and an energy loss arises. Loss is larger for larger relaxation time. Thus, the temperature dependence of α reflects the temperature dependence of the electron relaxation time. In the present temperature range, α is smaller in the deformed crystal. This is explained by considering that the electron relaxation time is shortened due to the scattering of electrons by dislocations. (d) Superconducting temperature (< 1.18 K). With decreasing temperature α steeply decreases. This is caused from the diminishing of the electron damping due to the decrease of the number of normal electrons. The study concerning this topic will be presented in the next section.

In the following, the sound attenuation due to electron damping is considered. Pippard has made a calculation based on a semi-classical theory on the electron-sound interaction [23, 24]. Using his formulation we can calculate the sound attenuation value, where the electron mean free path l is considered as a parameter. The mean free path is represented as

$$l^{-1} = l_i^{-1} + l_d^{-1}, \tag{25}$$

where l_i and l_d are mean free paths determined by the impurity scattering and by the dislocation scattering, respectively. In the present Al specimen of very high purity, experimentally determined value of l_i is about 1 mm at 4.2 K [25]. Meanwhile, l_d is written as

$$l_d^{-1} = d\Lambda, \tag{26}$$

where d is called the scattering cross-width of dislocation for electrons, and Λ is the dislocation density. Using the Pippard formula we obtain the value of d = 5.10 $\times 10^{-8}$ cm [22]. Here the dislocation density determined by the overdamped resonance data of IF is used. The d value obtained seems to be acceptable. Figure 20 shows how the derived value of d is altered when l_i is changed. For large enough l_i the value of d becomes constant. Thus the use of specimen with very high purity (large l_i) is important in such a study of electron-dislocation interaction.

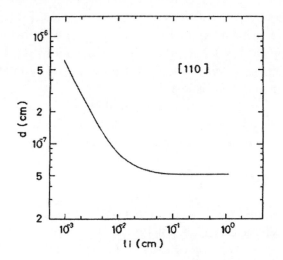

Figure 20. Change of scattering cross-width with impurity mean free path.

Electrical resistivity of metals is greatly affected by crystal defects. The resistivity due to dislocations is written as

$$\rho_d = P\Lambda, \tag{27}$$

where P is called the dislocation specific resistivity (DSR). The P value can be calculated using the value of d above obtained. In the present case of Al, the calculated value of DSR is shown to be $P = 2.03 \times 10^{-19}$ Ωcm^3. This value is well comparable with those obtained by other authors from the usual electric resistivity measurement: $P = (1\text{-}7) \times 10^{-19}$ Ωcm^3 [22].

H. INTERNAL FRICTION IN SUPERCONDUCTING ALUMINUM

In a superconductor a discontinuity in the electronic density of state, or the energy gap, exists near the Fermi surface. This has firstly been shown by the BCS theory [26] and then certified by a number of experiments. The energy gap can be anisotropic in real crystals, and study concerned is important for further development of precise theory of superconductivity. We have carried out such a sort of study [27]. Very pure (99.9999 % purity) and perfect (dislocation density $\approx 10^4\text{--}10^5$ cm^{-2}) Al crystals are used. The use of such crystals is shown to be useful in carrying out the present study. Crystals with three crystallographic orientations, [100], [110], and [111] are adopted. Ultrasonic pulse method is used for the attenuation measurement in the temperature range of 0.4–1.2 K, where ^3He pumping cryostat of usual type is utilized. Sound input power of the lowest level is used in order to avoid the effect of dislocation hysteresis loss.

Data of the temperature dependence of ultrasonic attenuation α are shown in figure 21 for three kinds of specimens. It has been shown by the BCS theory that the temperature dependence can be represented as

$$\alpha_s/\alpha_n = 2[1 + \exp(\Delta/k_B T)]^{-1}, \tag{28}$$

where α_s and α_n are sound attenuation in the superconducting and normal states, respectively, and Δ is the energy gap. An example of the BCS fit is shown in figure 22. We use essentially the same but somewhat modified formulation to determine the exact values of energy gap [27]. Note that the gap anisotropy we observed is for the specimen direction. Physically important is the gap anisotropy in the electronic k-space. After a precise argument we are able to determine the energy gap for any directions in the k-space [27]. In figure 23 the graphical representation of the gap anisotropy in the electronic k-space is shown, where the anisotropy is exaggerated five times. Examples of the gap values are

$$\Delta(100) = 3.56, \Delta(110) = 3.46, \Delta(111) = 3.14, \tag{29}$$

in the units of $(1/2)k_B T_c$, where T_c (= 1.18 K) is the transition temperature. The numerical values shown above are almost in accordance with the experimental and theoretical ones obtained by other authors [27].

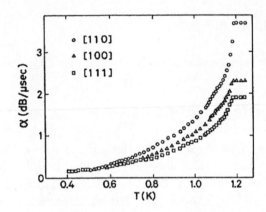

Figure 21. Temperature dependence of ultrasonic attenuation in specimens with different orientations.

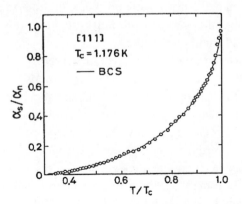

Figure 22. Normalized attenuation versus normalized temperature.

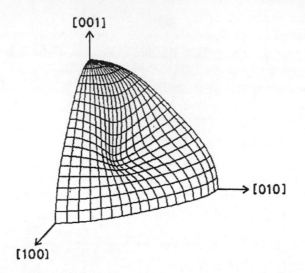

Figure 23. Graphical representation of spatial variation of energy gap in k-space.

I. HIGHER-ORDER ELASTIC CONSTANTS

In order to describe the nonlinear elasticity of materials in quantitative manner, we need values of the higher-order elastic constants. Complete sets of six third-order elastic constants (TOEC) have been determined experimentally for noble metals, Cu, Ag, and Au [28]. The measurements are carried out to obtain the values of TOEC from the changes of the second-order elastic constant (SOEC), or sound velocities, when the specimen is deformed by a uniaxial compression and a hydrostatic pressure. In order to avoid the effect of dislocation damping, the stress level must be very low, and a sensitive measurement method should be adopted.

A kind of ultrasonic interference method is used. An interference pattern due to sound pulses from two identical specimens is displayed on an oscilloscope. The first specimen is stressed, and the interference pattern is changed. Then, the temperature of the second specimen is varied to compensate the pattern change. The velocity change of the first specimen by the stress is obtained from the temperature change of the second specimen where the calibrated temperature dependence of sound velocity is utilized. Such a method is extremely sensitive to determine small amount of velocity change. The specimens used are 99.9995 % pure single crystals having (001),

(110), and (1$\bar{1}$0) faces and 15 × 16 × 17 mm in sizes. Specimens are pre-stressed by certain amount to increase the elastic range.

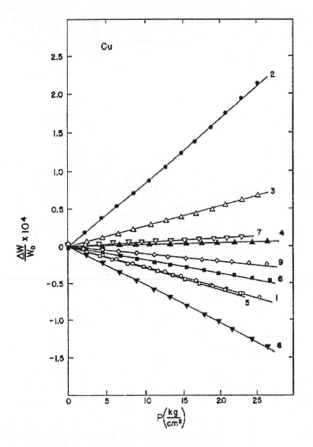

Figure 24. Velocity change with uniaxial stress in copper.

Examples of the experimental data for Cu are shown in figure 24 and figure 25 in the cases of application of uniaxial compression and hydrostatic pressure, P, respectively. Here, the quantity W is called the natural velocity, which is defined as twice the sound path length in the specimen in the reference (undeformed) state divided by the round-trip time of waves in the final (deformed) state. Note that only the natural velocity can be determined experimentally. We choose fourteen sets of measurements with various directions of the stress, sound propagation, and sound polarization. The numerals attached to the data in the figures indicate such identification. For example, No. 1 data: <001> compression, <110> propagation, and <110>

polarization; No. 10 data: hydrostatic compression, <001> propagation, and <001> polarization. Excellent linear relationship can be seen between the natural velocity change and the stress.

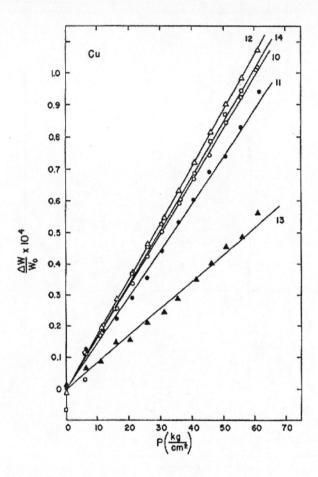

Figure 25. Velocity change with hydrostatic pressure in copper.

Such experimental data can be thoroughly analyzed for obtaining TOEC by the formulation given by Thurston and Brugger [29, 30]. In their formula, the natural velocity change is represented by a linear combination of SOEC and TOEC. The expression is rather complicated and will not be shown here (see [28]). We have fourteen sets of independent velocity change data for each material to determine six TOEC. Therefore, the determined values can be very reliable. The final results are shown in figure 26, where values of six TOEC, C_{IJK}, and three kinds of SOEC are plotted for Cu, Ag, and Au. Here, $B =$

$(1/2)(c_{11} + 2c_{12})$ is the bulk modulus, $C = c_{44}$ and $C' = (1/2)(c_{11} - c_{12})$ are the shear moduli. It can be seen that there are three categories of TOEC when their magnitudes are noted: large (C_{111}), medium (C_{112}, C_{166}) and small (C_{456}, C_{123}, C_{144}). Such results are interpreted on a basis of the atomistic argument. Elastic constants are determined by the interaction of atoms (ions) in the material. The above results can be explained by considering that the closed-shell repulsive interaction between nearest-neighbor atoms makes the dominant contribution to the higher-order elastic constants in the case of noble metals [28].

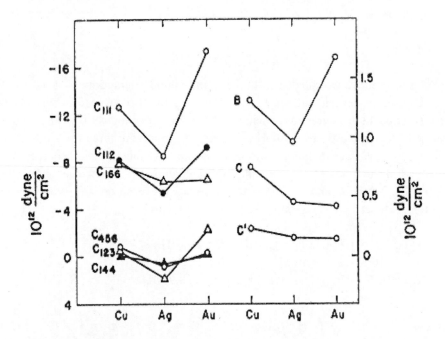

Figure 26. Second- and third-order elastic constants in copper, silver, and gold.

Here, our studies on the theoretical calculation concerning the higher-order elasticity are cited. The fourth-order elastic constants (FOEC) are evaluated, and the temperature dependences of SOEC, which are functions of SOEC, TOEC, and FOEC, are calculated [31]. By considering the crystal energy and its change by homogeneous deformation, TOEC and FOEC for Cu [32] and LiF, NaCl, and KCl [33] are calculated. The generation of the second harmonics of sound waves propagating in uniaxially stressed cubic crystals is discussed [34]. These studies are for the purpose of further understanding of the higher-order elasticity theory.

J. SOUND-SOUND INTERACTION

When a sound wave (1) and another sound wave (2) are collided in a material, a new sound wave (3) is generated. This is one of the typical phenomena caused from the nonlinear elasticity, or lattice anharmonicity, of materials. Such kind of phenomenon is conveniently described as a non-collinear three-phonon process where the conservation laws for momenta and energies hold:

$$q_1 + q_2 = q_3,$$
$$\omega_1 + \omega_2 = \omega_3. \tag{30}$$

From such an idea we can deal with the sound-sound interaction.

Ultrasonic experiment has been carried out for the purpose of observing the sound-sound interaction phenomenon [35]. Specimens used are zone-refined Cu single crystals of short cylindrical form having three flat side faces. On each of two faces a pulse transducer, and on one face a pulse receiver, are attached. Two types of experimental arrangements are adopted as shown in figure 27, where q and P represent the sound propagation and polarization vectors, respectively. The sound frequencies used are

Type I: $f_1 = 15$ MHz, $f_2 = 15$ MHz, $f_3 = 30$ MHz
or $f_1 = 10$ MHz, $f_2 = 10$ MHz, $f_3 = 20$ MHz,
Type II: $f_1 = 10$ MHz, $f_2 = 8.9$ MHz, $f_3 = 18.9$ MHz. (31)

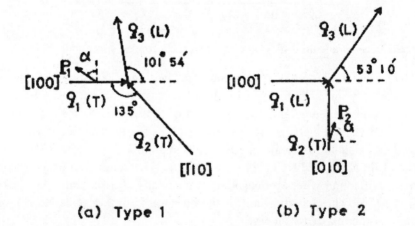

(a) Type 1   (b) Type 2

Figure 27. Geometry of three-phonon interactions.

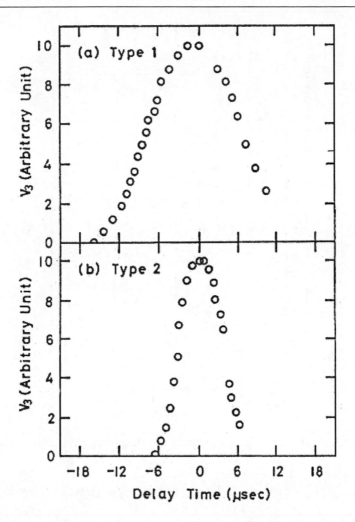

Figure 28. Change of received signal voltage when one of incident sound pulses is delayed against the other.

When one of the incident pulse wave (1) is delayed against another one (2), the received signal voltage V_3 changes as shown in figure 28. Namely, the sound-sound interaction strongly occurs when two incident pulses just collide in time. Figure 29 shows that the received signal voltage V_3 is proportional to the product of the voltages of the incident pulses, V_1 and V_2. This is in accordance with a conclusion of the three-phonon process theory. It is noted here that the collinear and non-collinear three-phonon processes can be used for determining the values of combinations of SOEC and TOEC.

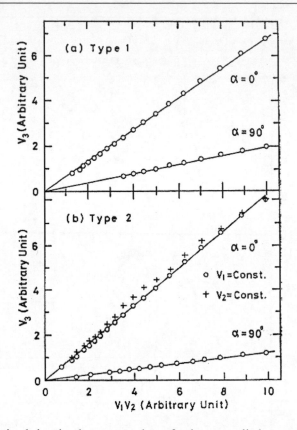

Figure 29. Received signal voltage vs product of voltages applied to two transducers.

K. NONLINEAR ELASTICITY IN WHISKERS

Whiskers are thin crystals with extremely high yield stress. The elastic ranges in the stress-strain relation are very wide in these crystals. Thus, we can expect to observe directly the elastic nonlinearity using the whisker crystals. The study for such a purpose has been performed [36] and the results were discussed using the results of the higher-order elasticity study [28].

Cu whisker specimens are grown by the reduction method. Namely, by the hydrogen reduction of a mixture of CuI and CuBr at 600°C for 2-3 h we obtain whisker crystals of [100], [110], and [111] crystallographic directions with sizes of 0.2-5 cm in length and 1-20 μm in diameter. The tensile apparatus adopted is a variant of an analytical balance where the load is applied by a solenoid-magnet system, and the displacement is detected by an optical lever system [37].

Typical stress P vs strain ε curves are shown in figure 30. Deviation from the linear elasticity (broken line) can be seen in the experimental data (solid line), and then a yielding occurs at a large stress. In the present case, the nonlinear stress-strain relation can be represented as

$$\varepsilon = P/E + \delta \, (P/E)^2, \tag{32}$$

where E is the Young's modulus and δ is called the nonlinearity constant. The value of constant δ can be determined from the experimental data. In figure 31 the experimental values δ_{exp} are plotted against the observed values of the yield stress P_{max}. There is a strong correlation between the two. Namely δ_{exp} approaches a constant small value at large P_{max}. Such a result is considered as follows. The stress in the specimen is composed of two parts, P_{e} and P_{i}. The former is the stress produced by the applied load, and the latter is the internal stress produced by some kind of defects in the specimen. When P_{i} is large, the yield stress P_{max} becomes small, while the nonlinearity constant δ_{exp} apparently becomes large. On the basis of such a consideration, an analysis is made where δ_{exp} values are modified to obtain the most probable δ values in perfect crystals [36].

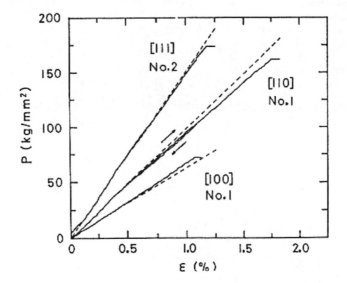

Figure 30. Stress vs strain curves of copper whiskers with different orientations.

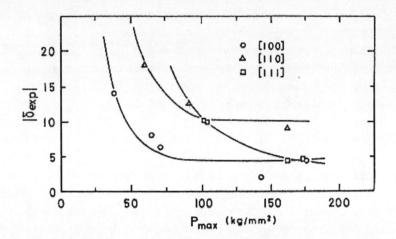

Figure 31. Experimental nonlinearity constant plotted against yield stress for three specimen orientations.

Meanwhile, the nonlinearity constant δ can be calculated by the nonlinear elasticity theory, and the constant is represented as a combination of SOEC and TOEC of the material [36]. We can evaluate the constant value using TOEC values of Cu determined previously [28]. In the following, the values of the most probable experimental values δ_{exp} and the theoretically calculated values δ_{cal} are shown for Cu crystals with different crystallographic orientations:

$$\delta_{exp}[100] = -4.3, \delta_{exp}[110] = 10.0, \delta_{exp}[111] = 3.5,$$
$$\delta_{cal}[100] = -4.5, \delta_{cal}[110] = 10.6, \delta_{cal}[111] = 3.8. \tag{33}$$

Excellent agreement between the experimental and the calculated values of the nonlinearity constants can be seen.

Chapter 4

RECENT RESEARCH

ABSTRACT

Shear viscosity of inorganic and organic glasses was measured near the glass transition. Temperature dependence of the viscosity was analyzed using a mechanical model to obtain the relaxation time. The hydrodynamic and hopping behaviors were shown at temperatures above and below the glass transition (A). Internal friction of polystyrene was measured at low frequencies at various temperatures. The results were analyzed using the Maxwell mechanical model, and the relaxation time for the hopping behavior was determined. Compensation effect relating the pre-exponential factor and the activation energy was seen (B). Internal friction was measured for various metallic glasses with widely different glass-forming ability. The increase of internal friction with temperature was considered on the basis of the idea of glass-forming ability. The compensation effect was considered using the idea of the complex relaxation representing simultaneous jump of relaxing elements (C). Isothermal annealing experiment using the internal friction measurement was performed for a metallic glass. The annealing behavior was of relaxation type. The experiments were carried out below and above the glass transition temperature. The relaxation was of the hydrodynamic- and hopping-type above and below the glass transition (D). Temperature, frequency, and amplitude dependences of internal friction were studied for a metallic glass. Internal friction peaks which shift with frequency were observed. The origins of the peaks were considered as the anelasticity due to stress-induced ordering. Severe fluctuations were observed in the amplitude dependence of internal friction near the glass transition, and the result was considered qualitatively (E). Ultrasonic and Brillouin scattering experiments were performed for polystyrene to obtain the sound velocity and attenuation.

The temperature dependence of sound velocity behaved as has been expected. From the sound attenuation data, apparent values of activation energy for relaxation were determined using various methods. The values were compared with each other, and also with the values obtained from the viscosity and internal friction studies. The idea of potential energy landscape was used for discussion (F).

A. VISCOSITY OF INORGANIC AND ORGANIC GLASSES; F~0 HZ

Various kinds of materials can be amorphous states, or glasses, under adequate conditions. In the glass-forming process, a material in the liquid state is cooled to the supercooled state instead of the crystalline state, and then to the glassy state at glass transition temperature T_g. The supercooled state is a disordered quasi-stable liquid state, and the glassy state is a frozen-in not-stable liquid. The energies of both states are higher than that of crystal and are apt to change to more stable values. Atoms in these states can move easily and relaxations due to the atomic motion occur. The relaxation can be studied by mechanical dynamical methods, and the methods are characterized by the adopted prove frequencies f. We have performed experimental studies of measuring shear viscosity $f \sim 0$ Hz; internal friction $f = 0.001$-10 Hz, ultrasonic attenuation $f = 5$ MHz; and Brillouin scattering $f = 10$ GHz. These studies will successively be presented in the present chapter.

Viscosity of inorganic glasses [38-43] such as metaphosphate (MP) and others, and amorphous polymers [41,43-45] such as polystyrene (PS), poly(methylmethacrylate) (PMMA), and polycarbonate (PC) has been measured near T_g. Viscosity η of glass-forming materials markedly depends on temperature. In the supercooled region, η value is very low at higher temperatures, and is gradually and then rapidly increased as the temperature is lowered. In the glassy state, η value is very large and still increases with decreasing temperature. We must adopt different experimental methods in the higher and lower temperature regions. The experimental method used at low-temperature high-viscosity region has been developed by us [40]. The method is a kind of sandwich method. Lower and upper faces of a cubic specimen are bonded to a fixed and movable test plate, respectively. A constant lateral load is applied to the movable plate by means of a weight and pulley. The lateral displacement of the plate is measured as a function of time by a laser measurement system. The temperature of the specimen is varied using a small cylindrical furnace set around the specimen. The deformation of the specimen is of almost pure shear mode, and the shear stress, the shear deformation, and the shear rate are in quite low levels. The viscosity measurements are carried out at successively higher temperatures from room temperature to a temperature close to T_g, and each measurement is made at a fixed temperature after the measurement system has sufficiently been stabilized. Using the apparatus we can measure the viscosity as high as 10^{13} Pa s. Meanwhile, the experimental method used at high-temperature low-viscosity region is as follows. The rotation disk viscometer method is adopted, and the apparatus used is the "dynamical analyzer" RDA-II (Rheometrics Inc., NJ, USA). This method is well established for measuring low viscosity (10^{-1}-10^8 Pa s) in a pure shear mode. We can directly obtain the experimental viscosity values by using the apparatus.

The method to obtain the η value in the low-temperature high-viscosity region is shown [40]. Typical data of lateral displacement of upper plate D vs time t are shown in figure 32. The experimental procedure is as follows. A load W is suddenly applied, an instantaneous increment of D occurs, and then gradual increase of D proceeds. Then the load W is suddenly removed, and instantaneous decrease of D occurs, gradual decrease of D proceeds, and D approaches a value higher than the initial zero value. The experimental load vs displacement relation is converted to the shear stress σ ($=W/S$, W: load, S: area of upper specimen surface) vs strain ε ($=D/H$; D: lateral displacement of upper specimen surface, H: specimen height). The observed ε-vs-t behavior can well be represented by the mechanical model shown in figure 33 (a). The composing elements of the model are elastic springs and viscous dashpots, and the right and left parts produce reversible anelastic and irreversible linear viscous behavior, respectively. The right part is called standard linear solids [46, 47]. Such an anelasticity-plus-viscosity model behaves as follows. When a constant stress σ is applied at time t=0 and then removed at t=t_0, the produced time-dependent strain ε(t) and ε'(t') are as shown in figure 33 (b). From the standard anelasticity theory [47], the produced strains are given as

$$t < t_0, \sigma \neq 0: \varepsilon(t) = a + b[1\text{-exp}(-t/\tau)] + ct, \qquad (34)$$

$$t > t_0, \sigma = 0: \varepsilon'(t') = d + b \exp(-t'/\tau), \qquad (35)$$

where a, b, c, and d are constant parameters, and τ is the relaxation time.

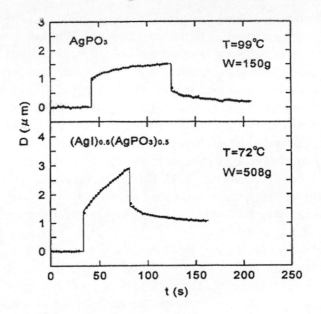

Figure 32. Displacement vs time for different glasses.

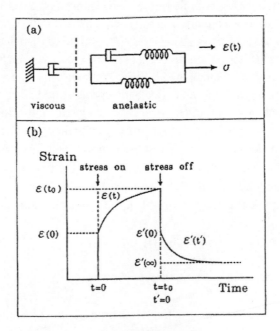

Figure 33. Mechanical model (a) and strain vs time (b).

By using the experimental $\varepsilon'(t')$ curve and Eq. (35), d, b, and τ are determined. Then, by using the experimental $\varepsilon(t)$ curve and Eq. (34), a and c are obtained. The solid lines in figure 32 are the parameter-fitted curves, and the fitting seems to be reasonable. The viscosity η is obtained in the loaded state ($\sigma \neq 0$) by definition:

$$\eta(\sigma \neq 0) = \sigma/[d\varepsilon/dt(t \to \infty)] = \sigma/c. \tag{36}$$

The viscosity η can also be determined in the unloaded state ($\sigma = 0$). The strain $\varepsilon'(t')$ finally approaches the remaining strain produced by the viscous flow during the loaded state, namely, $\varepsilon'(\infty) = d = ct_0$. Thus the viscosity can be obtained in the unloaded state as

$$\eta(\sigma = 0) = \sigma/[d/t_0]. \tag{37}$$

By using such methods of analysis, the viscosity can be separated from the anelasticity, and the viscosity value is more accurately determined in comparison with usual method in which the effect of anelasticity is ignored.

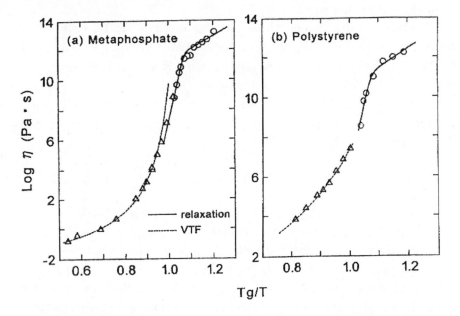

Figure 34. Temperature dependence of viscosity above and below glass transition measured by rotation method and sandwich method for metaphosphate glass (a) and polystyrene (b).

Examples of the results of viscosity measurement are presented in figure 34 [43]. The metaphosphate (MP) glasses, $(AgI)_x(AgPO_3)_{1-x}$; x=0.0-0.5, are made in our laboratory. The procedure of producing raw glassy materials is described in [42]. The raw material of PS is a thick large plate stored in our laboratory, which is considered to have been aged for a long period at room temperature. The molecular weights of PS are: number-averaged M_n=7.7×10^8. and weight-averaged M_w=18.1×10^8. The calorimetric glass transition temperature T_g is determined by the differential thermal analysis (DTA) [42]. The dependence of the viscosity η of MP and PS on temperature T is determined using the sandwich method from room temperature up to the glass transition T_g. At higher temperatures, the rotation disk method is used. In the case of polymers, the high-temperature viscosity apparently depends on the strain rate [48]. Thus, several viscosity measurements are performed with a decreasing strain rate, and the viscosity value in the limit of zero strain-rate is determined by an extrapolation. The results for MP (x=0.5) and PS are shown in figure 34, where the logarithm of the viscosity η is plotted against the inverse absolute temperature normalized by the glass transition temperature, T_g/T. Data points in the two temperature ranges obtained by different two methods relate quite smoothly.

The above experimental results are considered in the following. In the temperature range below T_g, the viscosity markedly increases with decreasing temperature and log η $\propto T^{-1}$. The rapid increase is gradually weakened in the lower temperature region, and the dependence finally comes again to be log η $\propto T^{-1}$. Such temperature dependence can be explained by considering a combination of two relaxations of thermal activation type, which is called double relaxation [42]:

$$\eta^{-1} = \eta_0^{-1}\exp(-E/RT) + \eta_0'^{-1}\exp(-E'/RT), \qquad (38)$$

where R is the gas constant. Let the first and the second terms represent the high-temperature relaxation (HTR) and the low-temperature relaxation (LTR). In figure 34 parameter-fitted curves for the two materials are shown by solid lines, where η_0, E etc. are chosen as fitting parameters. The results of the fit seem to be sufficiently acceptable. We consider that HTR appearing near and below T_g arises from the overall severe molecular motion in the glasses, since the activation energy for the relaxation is large. Meanwhile, LTR can be due to partial movement of groups of molecules in the molecular chains, since such relaxation occurs only in glasses containing long molecular chains [42].

Details of these topics are presented in our article [45]. Meanwhile, for the temperature dependence of viscosity above T_g, a formula named the Vogel-Tammann-Fulcher (VTF) equation [49] is usually used:

$$\eta = A \, \exp[B/(T\text{-}T_0)], \tag{39}$$

where A, B, and T_0 are constants. The broken curves in figure 34 are the result of the fit, and the fitting is agreeable. The VTF formula is originally an empirical expression. The formula, however, can be interpreted on an atomistic basis by the free volume theory [42, 49]. Finally, the overall feature of the temperature dependence of viscosity in glasses is shortly summarized. As temperature is lowered, the VTF fit at high temperatures (the hydrodynamic regime) gradually deviates from the data near T_g, and the fit of thermal activation type at low temperatures (the hopping regime) becomes preferable (crossover) [50].

B. INTERNAL FRICTION OF POLYSTYRENE; F=0.001~10 HZ

Internal friction (IF) measurement has been carried out for an organic glass polystyrene (PS) over a wide frequency range of f = 0.001-10 Hz and in a temperature range of 11-74°C [51, 52]. In the present IF experiment a method of "constant-temperature mechanical spectroscopy" is adopted. An inverted torsion pendulum composed of a thin-plate specimen is constructed, and a forced vibration of the specimen is induced by an AC magnetic excitation at a constant frequency. The phase difference δ between the external force and the induced deformation of the specimen is measured to determine IF of the specimen. The apparatus used is the Vibran Model MP-2001 with controlling and recording units (Vibran Technologies, Inc, MD, USA). The specifications of the apparatus are: frequency range 10^{-5}-10 Hz, frequency resolution 10^{-3}, strain amplitude 10^{-6}-10^{-4}, amplitude constancy 10^{-2}, specimen temperature 10-90°C, temperature constancy 10^{-3}. Values of IF are automatically measured as a function of frequency at various fixed temperatures and IF values are recorded on a chart. In the present method, an apparent mechanical loss occurs when the measurement frequency approaches the characteristic vibration frequency of the specimen. For the present specimen geometry such a loss does not occur for frequencies below 10 Hz. The PS specimen used is a thin plate of 75 × 8 × 1 mm in size with two holes for attaching the specimen to a specimen holder. The raw material is the same as used in the viscosity measurement. The glass transition temperature T_g of the specimen is determined by the differential scanning calorimetry (DSC) method, and the obtained value is $T_g = 88.0^{\circ}C$.

The measure of IF used is $Q^{-1} = \tan \delta$ where δ is the phase angle between the stress and the strain. The IF measurements are carried out in the frequency range of f=0.001-10 Hz, and 14 frequencies are chosen in each of the frequency ranges f=0.001-0.01, 0.01-0.1, 0.1-1.0, and 1.0-10 Hz, giving a high density of measurement points. Q^{-1}-vs-f measurements are performed at several fixed temperatures between $11^{\circ}C$ and $74^{\circ}C$, and 10 measurement temperatures are adopted. Smaller intervals for the measurement temperatures are adopted at higher temperature, where Q^{-1} changes rapidly with temperature.

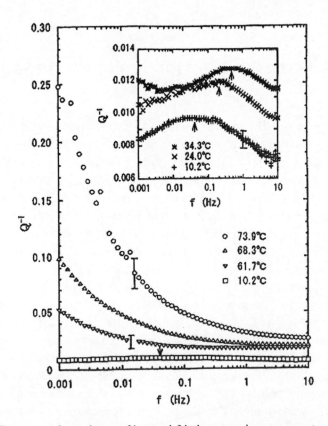

Figure 35. Frequency dependence of internal friction at various temperatures.

In figure 35 the overall behavior of Q^{-1}-vs-f is presented, where data at several temperatures are shown. The characteristics of the results are as follows. (a) IF increases when the measurement frequency is decreased, and the increase is larger at higher temperatures. (b) IF increases when the measurement temperature is increased, and the increase is larger at lower frequencies. (c) In addition, there exists a small IF peak as can be seen in the enlarged inset figure. In the following, we solely consider the items (a) and (b), since the IF peak (c) may be just a parasitic one.

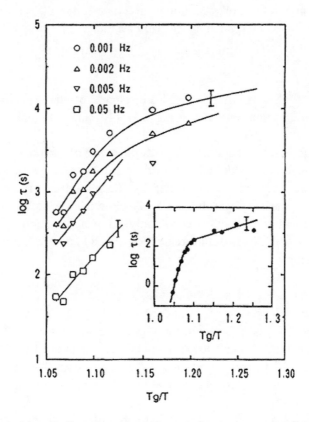

Figure 36. Logarithm of relaxation time vs inverse temperature normalized by the glass transition temperature at various frequencies.

We consider a mechanical model (Maxwell model) of a series connection of a spring (elastic modulus G) and a dashpot (viscosity η) [49]. The Maxwell model is primarily used for treating the viscous relaxation loss of liquid, and is frequently used for glass-forming materials since their characteristics are liquid-like both in the supercooled and in the glassy states [49]. We call such a kind of relaxation the "viscosity relaxation". IF of the Maxwell model arising from the viscous loss is given as [47, 53]

$$Q^{-1} = G/\eta\omega, \tag{40}$$

where ω is the angular frequency of vibration. The experimental results described previously as items (a) and (b) can thus be understood after considering that the viscosity η of glasses usually decreases when temperature is increased. Further, for a relaxation in a viscous medium the relaxation time τ is generally given by the following Maxwell relation [49]

$$\tau = \eta/G. \tag{41}$$

This relation is also extensively used in science of glasses [49]. We can determine the relaxation time τ by using Eqs. (40) and (41) and experimental Q^{-1} value.

We are interested in the temperature dependence of the relaxation time. In figure 36 the logarithm of the relaxation time τ obtained is plotted against the inverse absolute temperature normalized by the glass transition temperature, T_g/T, where the measurement frequency is chosen as a parameter. For the lower frequency data, the results can be analyzed using the "double relaxation" composed of the high temperature relaxation (HTR) and the low temperature relaxation (LTR):

$$\tau^{-1} = \tau_0^{-1}\exp(-E/k_BT) + \tau'_0{}^{-1}\exp(-E'/k_BT), \tag{42}$$

which is equivalent to Eq. (38). For the higher frequency case, reliable data at low temperatures are difficult to be obtained. The data are fitted only taking HTR into account. The solid curves in the figure represent the fitted ones. The inset figure shows the double relaxation obtained from shear viscosity measurement for PS.

We determine the pre-exponential factor τ_0 and the activation energy E for HTR at all measurement frequencies, and for LTR in a restricted frequency range. In figure 37 values of log τ_0 and log τ'_0 are plotted against E and E' for HTR (a) and LTR (b), respectively. An apparent linear relationship can be seen between the two quantities in both HTR and LTR. Such a correlation is called the "compensation effect", and has already been observed in the case of our viscosity experiments for inorganic [42] and organic [44] glasses. Such a correlation has firstly been observed in viscosity measurements of polymers by

Tanguy et al. [54], and was called the compensation effect. The meaning of this terminology is as follows: when the relation $\tau = \tau_0 \exp(E/k_B T)$ is considered, the decrease of the pre-exponential factor τ_0 and the increase of the activation energy E tend to hold the relaxation time τ constant. The compensation effect seems to be strange since the two quantities should primarily be independent each other when their physical meanings are considered. We have recognized that such an effect is frequently observed in various types of thermal activation phenomena in various kinds of materials [44, 45, 55]. The present author has developed a phenomenological theory to explain the compensation effect [55, 56]. Details of this problem will be presented in the following section.

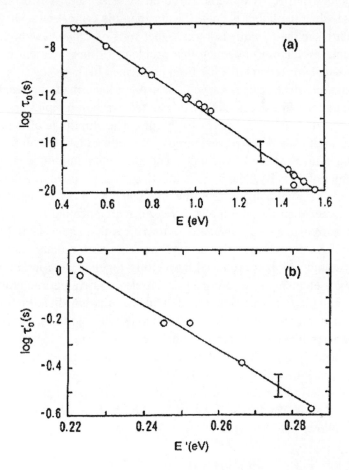

Figure 37. Compensation effect at high temperatures (a) and low temperatures (b).

C. Internal Friction of Various Metallic Glasses; f~1 Hz

Interest in metallic glasses (MG) is increasing because of their scientific and technological importance [57-60]. One of the important characteristics of MG is the glass-forming ability (GFA). MG with high GFA can be vitrified even when the material is cooled slowly from the liquid state. There are a variety of MG alloys, and GFA is largely different in different alloys. We can prepare glasses with large sizes for MGs with high GFA, and they are called the bulk metallic glasses (BMG). Most convenient measure of GFA is the critical cooling rate R_c. A material in liquid state can be transformed to the glassy state when the liquid is cooled faster than the critical rate. GFA is said to be higher when the R_c value is lower. What sorts of alloys have high GFA is an interesting problem. Concerning this problem, IF measurement of various MGs with widely different GFA has been performed [61].

The measurement of IF is carried out using a hand-made inverted torsion pendulum, and the free decay method is adopted. Ribbon samples are used and the gauge length of the specimen is 10-20 mm. The amplitude of the damping oscillation is optically detected, and three cycles of oscillation with frequency f are used for evaluating the IF value Q^{-1}. The maximum amplitude of the strain is of the order of 10^{-5}. By changing the length of the specimen and the moment of inertia of the oscillating system, the frequency f can be changed to some extent. The specimen temperature is increased with a heating rate of 1 K/min, and the measurements of Q^{-1} and f are performed with an interval of 3 K.

The method of preparing specimens is as follows. A mother alloy ingot of each alloy composition is produced by melting together appropriate amounts of constituent elements in an arc-melting furnace under an argon atmosphere. The mother alloy is melt-quenched with a single-roller melt-spinning apparatus in an argon atmosphere (4000 Pa) to produce a glassy ribbon sample of 10-30 µm in thickness, 1 mm in width, and several m in length. The melt-spinning speed is typically 3000 m/s. X-ray data are taken to confirm the amorphous state of the specimen. Binary, ternary, quaternary, and quinary MGs with widely different GFA are selected as the specimens. Adopted six MGs (No.1-6 alloys) and their R_c values reproduced from the existing data [59, 60] are as follows.

No.1 $Pd_{82}Si_{18}$: $Rc \sim 10^3$-10^4 (K/s),
No.2 $Pd_{77}Cu_6Si_{17}$: $Rc \sim 10^3$ (K/s),

No.3 $Pd_{40}Cu_{40}P_{20}$: $Rc \sim 10^2$-10^3 (K/s),
No.4 $Zr_{60}Al_{15}Ni_{25}$: $Rc \sim 10^1$-10^2 (K/s),
No.5 $Zr_{65}Al_{10}Ni_{10}Cu_{15}$: $Rc \sim 10^0$-10^1 (K/s),
No.6 $Zr_{41.2}Ti_{13.8}Cu_{12.5}Ni_{10.0}Be_{22.5}$: $Rc \sim 10^0$ (K/s).   (43)

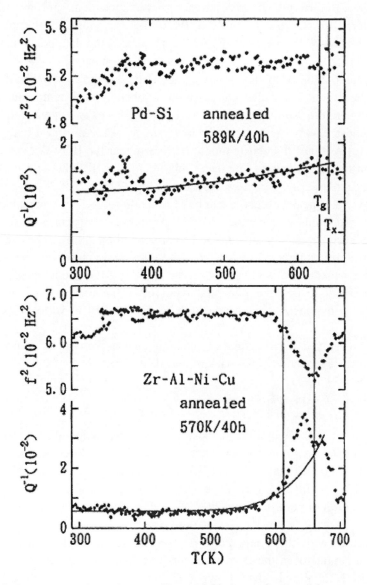

Figure 38. Temperature dependences of internal friction and squared frequency.

Typical data of the temperature dependences of the internal friction Q^{-1} and the value of f^2, which is proportional to the shear modulus of the specimen, are shown in figure 38 for No.1 alloy and No.5 alloy. The DSC values of the glass transition temperature T_g and the crystallization temperature T_x are indicated in the figure. Before the IF measurements the specimens are annealed at temperatures 40 K below T_g for 40 h. By such a treatment the specimens are stabilized, namely, the value of IF and the scatters of the data points are much reduced [62]. Experimental runs are repeated for each alloy to check the overall reproducibility of the data. Characteristics of the data are as follows: (a) The background IF value increases monotonously with temperature from room temperature, through T_g, up to T_x. (b) IF peaks superposed to the background are sometimes seen at a lower temperature and at a temperature near T_x. (c) A dip of the elastic modulus is seen near the IF peak temperature. (d) When the temperature is increased above T_x, the IF value and the elastic modulus decreases and increases, respectively.

Here, the IF peaks observed can be of relaxation type since the anomaly in the elastic modulus is accompanied. These peaks are structure-sensitive and material-sensitive, and are difficult to be considered on a unified basis [62-64]. Meanwhile, we can generally deal with the monotonic-increasing background IF. Such a background is always observed in other MGs we adopted, and the increase seems to be of exponential type [62-64]. Other authors have also observed the exponential increase of the background IF in various MGs [65-69].

It is presumed that in the experimental data as shown in figure 38 the temperature dependence of the background IF is given as

$$Q^{-1} = A + B \exp(-E/k_B T), \tag{44}$$

where A is an instrumental mechanical loss, B is the temperature-independent pre-exponential factor, and E is the activation energy. With choosing A, B, and E as parameters the data are fitted to the above equation. The fitting is made in the temperature range from room temperature up to T_x. The ranges where the IF peaks appear are omitted from the fitting. The fitted curves are shown in the figure.

Together with the temperature T, the measurement frequency f is an important parameter in the IF experiment. The frequency dependence of IF is studied for No.6 alloy as the following. Q^{-1}-vs-T measurements are carried out at various measurement frequencies f, the data are fitted to Eq. (44), and the parameter values are determined. The activation energy E is found to be almost independent of f, while there is a marked correlation between the pre-exponential factor B and the frequency f. The result is shown in figure 39, where B-vs-f^{-1} plot is given. In Eq. (44) only the second term representing the proper IF is taken. Then, for the IF the following relation holds:

$$Q^{-1} \propto f^{-1}. \tag{45}$$

Using such a procedure we can accurately observe the frequency dependence of IF. The frequency dependence is the same for the as-prepared and the annealed specimens. Namely, the frequency dependence is an essential characteristic of the material. Incidentally, the direct measurement of frequency dependence of IF with keeping the temperature constant is not so easy because of technical reasons.

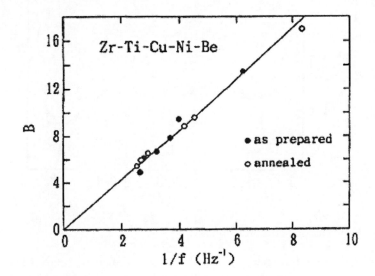

Figure 39. Pre-exponential factor plotted against inverse frequency for No. 6 alloy.

The frequency dependence of IF is further considered. We here take the Maxwell model, and IF arising from the viscous loss is represented by Eq. (40). The observed frequency dependence of IF, Eq. (45), is in accordance with that in the viscosity relaxation represented by Eq. (40). Further, the Maxwell relation Eq. (41) is adopted. Then by using Eq. (44) (omitting parameter A), and Eqs. (40) and (41) the relaxation time is represented as

$$\tau = (1/2\pi f B)\exp(E/k_B T) \equiv \tau_0 \exp(E/k_B T). \tag{46}$$

The activation energy E for the relaxation is determined using Eq. (44). The pre-exponential factor of the relaxation time τ_0 can be determined using the experimental B value and the measurement frequency f. In figure 40, $\log\tau_0$ are plotted against E for six alloys No.1-6, where the numerals attached to the data points represent the alloy numbers. Discussion on this result is made as follows.

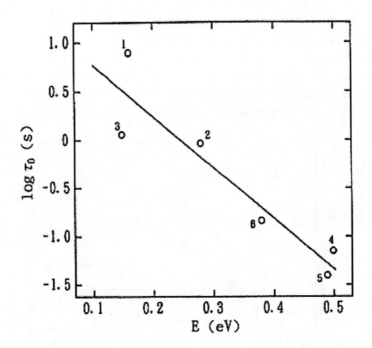

Figure 40. Compensation effect for six alloys.

First, the activation energy E is considered. As can be seen in figure 40, there is a correlation between the R_c value representing the GFA (see Eq. (43)) and the activation energy value E. Namely, when R_c is lower, or GFA is higher, E tends to be higher. Now, consider a glass with high GFA. The material in the liquid state is easily transformed to the glassy state when it is cooled down. Meanwhile, the material in the glassy state may not easily be transformed to the crystalline state when it is heated up. These features are due to the fact that in glasses with high GFA the glassy state is much stable. Here, consider the course of the heating. Partial crystallization is considered to progress gradually over a wide temperature range below the crystallization temperature T_x. The crystallization occurs more actively as the temperature is increased up to T_x, and is finished at T_x. Such a process can be of thermal-activation type, and the activation energy represents the averaged height of the energy barrier for the process. Thus, the GFA of MG is higher, the barrier for the activation process is higher, and the observed activation energy value E is larger. It is also added here that the E value is much different in specimens in different states. For example, for No.1 alloy, in as-prepared specimen E=1.05 eV and this value is very large compared with that in the annealed specimen E=0.16 eV (see figure 40). The activation energy should be smaller in well-annealed specimen since the energy barrier for the activation process can be lowered by the annealing.

Second, an apparent linear relationship is seen between $\log\tau_0$ and E. Namely, the compensation effect previously mentioned can be seen also in the present case. It is shown [55. 56] that two kinds of relaxations of thermal-activation type exist: the simple relaxation (SR) and the complex relaxation (CR). In a thermal-activation process, relaxing elements jump over an energy barrier. In SR the jumps of relaxing elements occur independently. This is the relaxation usually observed, and the activation energy E and the pre-exponential factor τ_0 are independent each other. The observed pre-exponential factor is close to the value expected from the jump of a relaxing element. Meanwhile, in CR jumps of several relaxing elements occur simultaneously and collectively. The activation energy E and the pre- exponential factor τ_0 are correlated each other, and the compensation effect is seen between the two. The observed value of pre-exponential factor is anomalously larger than that expected. Two kinds of relaxation, SR and CR, are really observed in various relaxation phenomena [55, 56]. The present author has developed a phenomenological theory concerned [55, 56], which is described in the following.

Consider a SR, and let the number of successful trials per time for surmounting the energy barrier E is f_r. Next, consider a collective jump of n relaxing elements occurring simultaneously and independently (CR). The number of successful jumps after a time t_0 is $(f_r t_0)^n$. Here, t_0 is an indefinite parameter. The number of such successful jumps per unit time is $(f_r t_0)^n / t_0$. Meanwhile, $f_r = [\tau_0 \exp(E/k_B T)]^{-1}$, where E and τ_0 are the activation energy and the pre-exponential factor in SR composing CR. Therefore, the relaxation time for CR can be written as

$$\tau(n) = [(f_r t_0)^n / t_0]^{-1} = t_0^{1-n} \tau_0^n \exp(nE/k_B T). \tag{47}$$

The activation energy and the pre-exponential factor experimentally observed in CR are represented as E^{obs} and τ_o^{obs}, and they are given as

$$E^{obs} = nE, \ \tau_o^{obs} = t_0^{1-n}\tau_o^{n}. \tag{48}$$

Thus, as shown in the first formula, E^{obs} is always larger than E. In the second formula, let $t_0 = N\tau_o$ where $N>1$. Then, $\tau_o^{obs} = N^{1-n}\tau_o$ and in this case τ_o^{obs} is larger than τ_o. Namely, the relation $t_0>\tau_o$ is the condition required for observing $\tau_o^{obs}>\tau_o$. Now, from two formulas in Eq. (48) we obtain the expressions

$$\log\tau_o^{obs} = A_1 + A_2 E^{obs};$$
$$A_1 = (1-n)\log t_o, \ A_2 = \log\tau_o/E. \tag{49}$$

When A_1 and A_2 are constants, then the linear relation between $\log \tau_o^{obs}$ and E^{obs}, namely the compensation effect, is realized in a series of relaxation experiments.

Now the present experimental result for MGs shown in figure 40 is reconsidered. Here for τ_o and E in the figure, read τ_o^{obs} and E^{obs}. First, the observed relaxation time τ_o^{obs} is considered to be very large. Second, there is a linear relation between $\log\tau_o^{obs}$ and E^{obs}. Namely, the compensation effect is occurring in the present series of experiments, and here A_1 and A_2 in Eq. (49) should be constants. In the series of experiments using different MGs, when for each experiment the number n is nearly the same and nearly the same value of t_0 can be selected, then A_1 is expected to be a constant. Meanwhile, A_2 is constant means that the ratio of $\log\tau_o$ and E is constant for different MGs. This means that, in SR, when the activation energy peak is higher, the shape of the peak is broader since the trial frequency is smaller. This seems to be a characteristic of the SR occurring in the present case.

Some numerical evaluations concerning the present experimental result are given. As described before, the time t_0 is to be larger than the relaxation time τ_o in order to realize the characteristic large τ_o^{obs}. Using the formulas given in Eqs (48) and (49), we obtain the relations

$$\log t_0 = A_1/(1-n), \ \log\tau_o = A_2 E^{obs}/n. \tag{50}$$

From these formulas and the expected inequality $\log t_0>\log \tau_o$, we obtain the following relation which restricts the number n:

$$n > (A_2 E^{obs})/(A_1 + A_2 E^{obs}). \tag{51}$$

The experimentally determined constant values are A_1 =1.30 and A_2 = -5.21, and the averaged value of observed activation energies is $<E^{obs}>$=0.33 eV. Thus we obtain the relation $n > 4.10$. Now, the value $n = 5$ is tentatively taken since the thermal activation process with much larger n value is not expected to be realized. Then, from Eq. (50), the value $<\log\tau_o>$=-0.34 is obtained. This value is very large compared with the value in the case of jumping single atom ($\log\tau_o \sim -14$). Namely, the frequency of vibration of the relaxing element in SR is very low. Generally, the frequency of a vibrating body is represented as $(F/M)^{1/2}$ where F is the force acting on the body and M is the mass of the body. Thus, in the present case, the relaxing element in SR can be a group composed of a number of atoms, and the force acting on the element can be rather weak. Weak force usually corresponds to a weak potential-energy field, and the observed activation energy can be low. Really, the observed activation energy is $<E^{obs}>$=0.33 eV, and the activation energy for SR is $<E>$=0.33/5 =0.066 eV. This value looks to be in a quite low level. In conclusion, it is considered that in the relaxation presently concerned large relaxing elements move collectively in weak potential-energy fields.

D. INTERNAL FRICTION OF METALLIC GLASS; F~1 HZ

The energies of supercooled and glassy states of glass-forming materials are higher than that of crystalline state, and energy decrease or "stabilization" easily occurs. Isothermal annealing experiments are performed to investigate the stabilization phenomenon near the glass transition temperature T_g [70, 71]. We are interested in mechanical relaxation in glassy materials, especially near T_g. The motivation is to understand the characteristics of glassy state and the glass transition phenomenon. The internal friction Q^{-1} and the oscillation frequency f are measured by the inverted pendulum method during the stabilization process. The specimen material used is the quinary alloy $Zr_{41.2}Ti_{13.8}Cu_{12.5}Ni_{10.0}Be_{22.5}$ having high glass-forming ability. A single-roller melt-spinning method is used for preparing the specimens. DSC values for the specimen are $T_g = 621.8$ K and $T_x = 712.9$ K.

Temperature dependent mechanical relaxation of the specimen is firstly shown. Figure 41 gives examples of the temperature dependences of internal friction Q^{-1} and squared frequency f^2. Measurement is performed from the room temperature, through T_g, and up to or somewhat above T_x. A heating rate of 1 K/min is adopted, and the data sampling is made in every 3 K. The features of the results are as follows. (a) Annealed at 620 K for 20 h. The data show the regular feature of a well-stabilized specimen annealed for a long time at a temperature near and below T_g [70]. The low-temperature IF value is almost constant and not much scattered. The value gradually and then rapidly increases with temperature very smoothly. Clear multiple-peak behavior is seen in the crystallization region. A slight anomaly (slope change) can be seen near T_g as indicated by an arrow in the figure (see [72]). The value of f^2 also regularly changes with temperature in accordance with the change of Q^{-1}. (b) Annealed at 670 K for 20 h. The Q^{-1} behavior becomes to be irregular. The high-temperature peak is ambiguous, and an unexpected small peak appears at a lower temperature. This result possibly shows that a partial crystallization, which is often observed in BMGs [57, 58], occurs in the material through annealing at a temperature above T_g and below T_x.

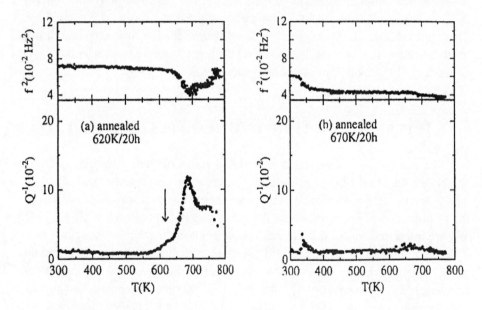

Figure 41. Temperature dependences of internal friction and squared frequency in different states after different annealing.

Now, an as-prepared specimen is annealed at a constant temperature T_a, and the changes of the internal friction and elastic modulus with time t are observed. The measurements are performed at several annealing temperatures in the range of $T_a = 615$-695 K, namely, T_a is near and below or above T_g. In every annealing experiment a new specimen is used. Figure 42 shows an example of the results. The temperature T is increased at a rate of 3 K/min from the room temperature up to T_a, and is kept constant. In the figure, the time t represents the time elapsed after the start of the experiment. After $t = t_0$ the temperature T becomes almost constant, and Q^{-1} monotonically decreases with time. In the experiment the measurement is made in every 3 min. The decrease of Q^{-1} is considered to represent the stabilization process through the annealing.

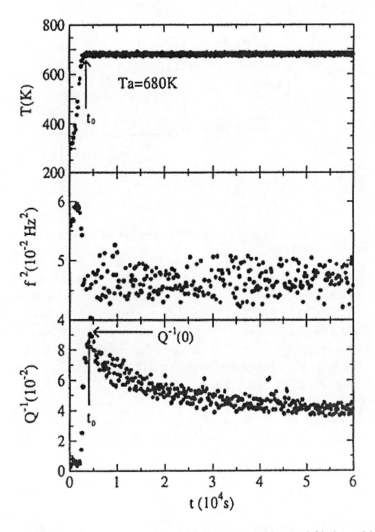

Figure 42. Changes of temperature, squared frequency, and internal friction with time after start of experiment.

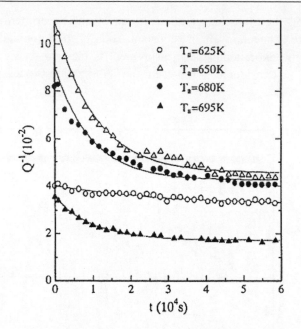

Figure 43. Analysis of stabilization process for several annealing temperature to determine the relaxation time.

In figure 43 several examples of the change of Q^{-1} with time are shown for different annealing temperatures T_a, and hereafter t represents the time elapsed after the time t_0. Let $Q^{-1} = Q^{-1}(0)$ at $t = 0$, and from this value Q^{-1} monotonically decreases with t. The data points shown in the figure are obtained after adopting an averaging process. Ten neighboring raw data points as shown in figure 42 are averaged to obtain a data point in figure 43 in order to reduce the scatter of data. The data as shown in figure 43 are used for analyzing the stabilization kinetics.

For analyzing the stabilization process, a relaxation with a single relaxation time is considered. Then Q^{-1}-vs-t can be represented by the formula [49]:

$$Q^{-1}(t) = Q^{-1}(\infty) + [Q^{-1}(0) - Q^{-1}(\infty)]\exp(-t/\tau), \qquad (52)$$

where $Q^{-1}(0)$ and $Q^{-1}(\infty)$ represent the initial and the final (equilibrium) values of Q^{-1}, respectively, and τ is the relaxation time for the stabilization process. By adopting the experimental value of $Q^{-1}(0)$ and adjusting the values of $Q^{-1}(\infty)$ and τ as parameters, Eq. (52) is fitted to the experimental data. The curves in the figure represent the fitted ones. In this manner the relaxation time τ is determined, and in figure 44 the values of logarithm of τ are plotted against the inverse of annealing temperature T_a^{-1}. It is interesting to see that the feature of $\log\tau$-vs-T_a^{-1} is different in the low, middle and high temperature regions.

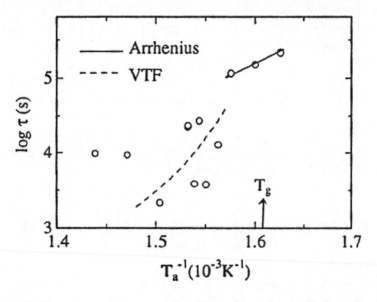

Figure 44. Logarithm of relaxation time vs inverse of annealing temperature.

The above result of $\log\tau$-vs-T_a^{-1} is now considered. The characteristics of the viscoelastic relaxation in glass-forming materials are, as described previously, summarized as follows. In the range of temperature T above T_g, the viscoelastic property of the material is liquid-like (hydrodynamic regime) for which the Vogel-Tammann-Fulcher relation is applied for the relaxation time:

$$\tau = A \exp[B/(T - T_0)]. \tag{53}$$

At lower temperatures the viscoelasticity of thermal-activation type showing the Arrhenius relation is preferable (hopping regime):

$$\tau = \tau_0 \exp(E/k_B T). \tag{54}$$

Above features are seen in the present $\log\tau$-vs-T_a^{-1} plot in the range of T_a^{-1} >1.5. Namely, the data in figure 44 in the hydrodynamic regime (T_a^{-1}>1.5) can be fitted to the VTF formula represented by Eq. (53). The broken line shows the fitted curve. The data scattering looks to be not so small. The scatter can primarily be due to experimental origins. However, it is also possible that the scatter is originated from the nature of the relaxation itself in this temperature range. The data in figure 44 in the hopping regime are fitted to the Arrhenius formula represented by Eq. (54). The scatter of the data is slight in this temperature range, and the solid line shows the fitted one. The transition between the above two kinds of behaviors seems to occur at a temperature near and somewhat above T_g. The data in the range of $T_a^{-1} < 1.5$ are finally considered. Effect of the partial crystallization occurring in this temperature range (see figure 41(b)) could be the origin of the deviation of the data from the VTF formula. Other authors also observed the same tendency, the enhancement of the viscoelastic relaxation time by crystallization [73]. Final conclusion is that the stabilization phenomenon can well be treated on the basis of the viscoelastic relaxation process.

E. INTERNAL FRICTION OF METALLIC GLASS; F=0.2~25 Hz

Temperature, frequency, and amplitude dependences of IF have been studied for the metallic glass $Zr_{41.2}Ti_{13.8}Cu_{12.5}Ni_{10.0}Be_{22.5}$ [74]. Measurement is performed by using "dynamical mechanical analyzer" (DMA, type Q800, TA Instrument, Delaware, USA). Value of IF can be obtained in the temperature range -150 - $600°C$, in the frequency range 0.01-200 Hz, and in the specimen displacement amplitude range 0.5-10000 μm. The deformation mode of the specimen adopted is cantilever bending. At a definite temperature with definite strain amplitude of vibration, internal friction and Young's modulus can quickly be measured at several preset vibration frequencies. In the experiment, bulk specimens 17.5 mm in length, 3.0-4.1 mm in width, 0.93-1.01 mm in thickness are used. DSC values of the specimen are $T_g = 621.8$ K and $T_x = 712.9$ K.

Typical data of temperature dependences of internal friction Q^{-1} and Young's modulus E measured at several frequencies are shown in figures 45 and 46 for an as-prepared specimen and for a specimen annealed at 620 K for 20 h. The measurement is made with increasing temperature at a rate of 1 K/min, and the data are taken every 3 K. The strain amplitude of vibration is fixed at 10^{-5}. The characteristics of the results are as follows. (a) As prepared: At a low temperature, a small peak is observed in Q^{-1}-vs-T, and a dip in E-vs-T is observed near the Q^{-1} peak position. A large Q^{-1} peak and a large E dip appear at a high temperature. It is seen that the position of the Q^{-1} peak indicated by the arrow shifts to a higher temperature with increasing measurement frequency. (b) Annealed at 620 K/20 h: The previously observed low-temperature Q^{-1} peak cannot be seen. Such a Q^{-1} peak may be due to an unstable state of the as-prepared specimen. The high-temperature Q^{-1} peak apparently exists. In addition, a peak can be observed in the medium-temperature range as indicated by the arrow. The peak is not so high, and can be observed in this specimen state having reduced background IF by the annealing. The position of the peak also shifts to a higher temperature with increasing measurement frequency. In figure 47, the logarithm of the experimental relaxation time, $\tau = 1/2\pi f$, is plotted against the inverse of the peak temperature T_p^{-1} for the high-temperature peak (a) and for the medium-temperature peak (b). Apparent correlations can be seen between the above two quantities.

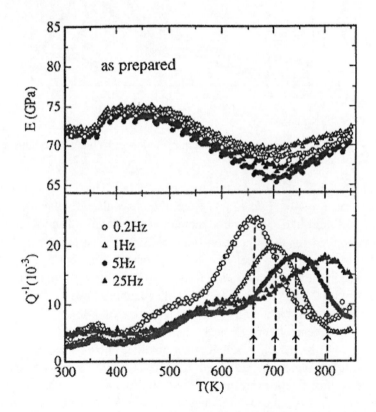

Figure 45. Temperature dependences of internal friction and Young's modulus for four vibration frequencies in as-prepared specimen.

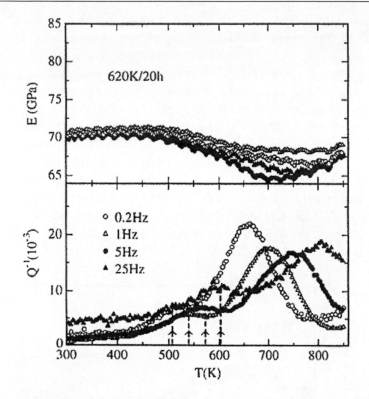

Figure 46. Temperature dependences of internal friction and Young's modulus in specimens annealed at 620 K for 20h.

The high-temperature IF peak is now considered. This peak is observed in the supercooled region ($T_p > T_g$). Here the viscoelastic relaxation in the material is liquid-like, or hydrodynamic, and Vogel- Tammann-Fulcher relation is applied for the relaxation time (Eq. (53)). The $\log\tau$-vs-T_p^{-1} data in figure 47 (a) are fitted to the VTF formula where T is to be read T_p. The broken curve in the figure shows the fitted one. The fitted parameters are $\log A$ = -10.3, B = 13025 K, and T_0 = 108 K. Other authors have carried out an IF study similar to ours for the alloy $Zr_{46.75}Ti_{8.25}Cu_{7.5}Ni_{10.0}Be_{27.5}$ [75]. They also observed a high-temperature IF peak, and from the peak shift the parameter values B = 9440 K and T_0 = 352 K were obtained. From these two kinds of results, the following points are noted. (a) The obtained two sets of parameter values are very different even though the tested materials are similar in their compositions. This is not so well understood when the IF peak is considered to be due to the hydrodynamic viscous relaxation, since the origin of which is the overall atomic motions of atoms in the material. (b) In both experiments, the obtained T_0 value seems to be unreasonable. At this temperature, the relaxation time diverges (see Eq. (53)), the hydrodynamic behavior cannot exist above this temperature, and thus the T_0 value should not be lower than T_g (~ 620 K). After considering these two things, the VTF formalism seems to be unsuitable in the present case.

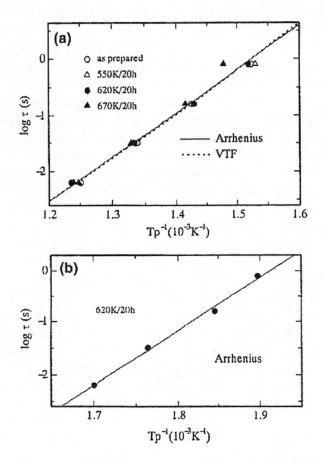

Figure 47. Analysis of high-temperature (a) and medium-temperature (b) internal friction peaks.

Now both the high-temperature and the medium-temperature IF peaks are considered to be due to relaxations associated with local motions of atoms or defects in the material. Furthermore, the motion is assumed to be of the thermal-activation type. Then we can adopt the Arrhenius-type relaxation time (Eq. 54)). Results of data fittings on this basis are shown by the solid lines in figures 47 (a) and (b). The fitted parameters obtained are

high-temperature peak: $\log\tau_0 = -11.9$, $E = 1.55$ eV,
medium-temperature peak: $\log\tau_0 = -19.8$, $E = 2.05$ eV.   (55)

It is well known that even in glassy materials, the Snoek-type anelasticity can appear. For example, a low-frequency IF peak is observed in hydrogenated MG at a temperature lower than room temperature [76]. Hydrogen atoms are often contained in usual MG specimens due to technical reasons, and a low-temperature H-peak is frequently observed. However, the H-peak always appears at low temperatures, and the IF peaks observed in the present case cannot be such H-peak. Now, in the case of the high-temperature peak, the value of $\log\tau_0$ nearly coincides with that for the hopping of a single atom, $\log\tau_0 \sim -12$ (Debye frequency $\sim 10^{12}$ Hz). Namely, the element contributing to the anelasticity may be a single atom, and the stress-induced ordering of atoms [47] can be the origin of the anelasticity. In the present material, the Be atom is the smallest and can move most easily. There have been studies of the diffusion of Be in the material using the backscattering technique [77, 78], and the obtained activation energy for the diffusion is 1.1 eV [78]. From the present IF result it is considered that the high-temperature IF peak can be related to the motion of Be atoms. Next, the case of the medium-temperature IF peak is considered. Obtained value of $\log\tau_0$ is smaller and the value of E is larger than those for the high-temperature IF peak. Thus, as described previously, the IF peak is considered to be due to a kind of complex relaxation. For example, for a group of Be atoms, which is the most probable candidate, the observed activation energy must be larger than 1.55×2 (atomic pair) = 3.1 eV. The observed activation energy value (2.05 eV) is meaningfully smaller than this. The origin of this IF peak is presently not certain.

The dependence of IF on the strain amplitude of vibration is studied for the as-prepared specimen with measurement frequency 1 Hz at various temperatures. The temperature is increased with a rate of 1 K/min to a certain temperature and kept constant. IF is measured as the strain amplitude A is increased. The amplitude range is $A=10^{-5}-10^{-4}$, and a measurement run is made within 30 min. The temperature is increased and the measurement is repeated. The measurements are carried out at 300 K, 550 K, $\cdots\cdots$, 800 K, and then again at lower temperatures 590 K and 640 K. The experimental results are shown in figure 48.

The above results are considered in the following. When a hysteresis exists between the applied stress and the induced strain in a material, a static mechanical loss proportional to the amount of the hysteresis occurs. IF due to the hysteresis occurs when a vibration stress is applied. The Granato-Lücke pinned-dislocation model is a typical example of this kind of loss [4]. In such hysteresis loss cases IF apparently depends on the strain amplitude. In the present results shown in figure 48, no definite overall increase or decrease of IF is seen (T=300, 550 K). However, fluctuations are seen in Q^{-1}-vs-A, which are especially remarkable at T = 620 and 670 K. These temperatures are near the glass transition T_g (~ 620 K). After the temperature is increased to 800 K, the fluctuations are apparently diminished even at temperatures near T_g (T = 590, 640 K).

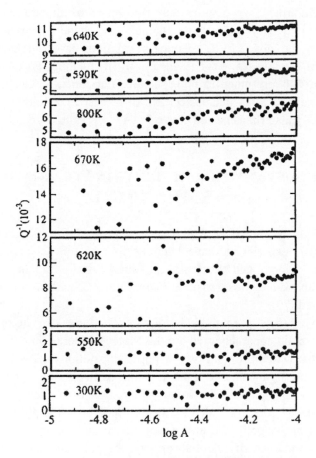

Figure 48. Amplitude dependence of internal friction at different temperatures.

Experimental results are tentatively explained as follows. Movable elements - atoms or groups of atoms - exist in the glasses, which can be presumed from the relaxation peaks observed in Q^{-1}-vs-T as shown previously. In the present case, in the unstressed state an element is in the equilibrium position at the bottom of the potential due to the surrounding elements. When an external stress is applied the element in the potential moves up and reaches an additional quasi-equilibrium small potential-minimum position. Rapid vibration of the element around this position is initiated by the applied static stress. The produced additional vibrating strain superposes to the static external strain. There may be a number of such quasi-equilibrium potential minima, and the superposition mentioned above occurs frequently in a stress-strain cycle of the IF measurement. Then a hysteresis occurs in the stress-strain cycle. Further, we assume that the shape of the quasi-equilibrium potential is slowly varying with time, namely, the shape is occasionally steep or flat. Then the observed Q^{-1} value is small or large in each Q^{-1}-vs-A measurement. Namely, Q^{-1}-vs-A is apparently fluctuating. Furthermore, near T_g the potential is presumed to be especially flat. Then IF can take notably large values near T_g. Such an enhancement of IF near T_g should be disappeared when the glassy material is heated above T_x (~713 K) and is crystallized.

F. ACOUSTIC MEASUREMENTS IN POLYSTYRENE; F = 5 MHz, 10 GHz

Two kinds of acoustic experiments, ultrasonic (US) and Brillouin scattering (BS) experiment, have been performed for polystyrene (PS) to determine the velocity and attenuation of sound from room temperature up to and slightly above the glass transition temperature T_g, and two kinds of results were compared [79].

In US experiment, the pulse echo method is adopted for the measurement [80, 81]. The main apparatus used are the Matec Model 6600 comparator unit and the Iwatsu DS8606 digital oscilloscope. Separations between echoes are measured using a time marker, and together with the specimen length the sound velocity is obtained. The sound attenuation is measured by fitting an exponential decay curve provided by the comparator to the pulse echoes. The accuracies of the velocity and attenuation measurements are around 10^{-3}. To measure the sound velocity more accurately, the pulse overlap method having an accuracy of 10^{-5} [80, 81] is adopted. Measurements are carried out for longitudinal (L) and transverse (T) sound waves. The specimen used has two 10×10 mm flat and parallel surfaces separated by several mm. The PS material used is the same as before. An L- or T-mode PZT ceramic ultrasonic transducer is used, and is bonded to a surface of the specimen with an adequate bonding material.

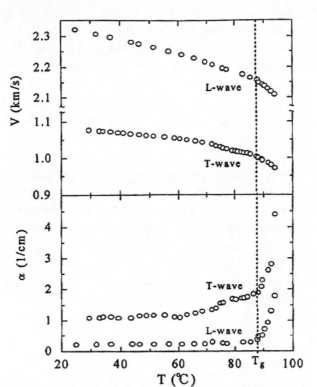

Figure 49. Temperature dependences of sound velocity and sound attenuation for L-waves and T-waves.

Typical results of the temperature dependences of velocity V and attenuation α in the cases of L- and T-waves are shown in figure 49. The glass transition temperature (= 88.0°C) is shown by the broken line. The characteristics of the sound velocity data are summarized as follows. (a) The velocity of L-waves is higher than that of T-waves. (b) The velocities of both L- and T-waves decrease with increasing temperature. The decrease is monotonic at low temperatures, and the decrease is steeper for L-waves. (c) The velocity decrease becomes more rapid near and above T_g for both L- and T-waves. No discontinuity of the velocity value is seen at T_g. The sound velocities in an isotropic medium are represented as [81]

L-wave: $V = [\{(4/3)G + K\}/\rho]^{1/2}$,

T-wave: $V = [G/\rho]^{1/2}$,   (56)

where G and K are the shear and bulk moduli, respectively, and ρ is the
material density. As can be seen from Eq. (56), the velocity of the L-waves is
higher than that of T-waves, which is observed in figure 49. The monotonic
decrease of the sound velocity with increasing temperature arises from the
decrease of the elastic modulus due to the anharmonicity of the material [81].
Thus, the decrease in velocity with temperature should be greater for L-waves
than for T-waves, which is as observed in figure 49. The more rapid decrease
of the velocity observed near and above T_g can be an effect of a kind of
mechanical relaxation.

The results of the sound attenuation seen in figures 49 are summarized. (a)
The attenuation of L-waves is smaller than that of T-waves. (b) The
attenuation of L-waves is almost constant at lower temperatures and increases
very rapidly near and above T_g. No discontinuity of the attenuation value is
seen at T_g. (c) The attenuation of T-waves gradually increases with
temperature, and the increase becomes rapid near T_g. No discontinuity of the
attenuation value is seen at T_g. Now, the origin of the sound attenuation is
considered to be due to a kind of mechanical relaxation. The sound attenuation
α and the internal friction Q^{-1} are related as [81]

$\alpha = Q^{-1} f/V$.   (57)

Using Eqs. (40), (41), (46), and (57) we obtain the following relation:

$\alpha = (2\pi V\tau_0)^{-1}\exp(-E/k_BT) \equiv B\exp(-E/k_BT)$.   (58)

The data in figure 49 are fitted to the formula

$\alpha = A + B\exp(-E/k_BT)$,   (59)

where the constant A represents instrumental background loss. Regarding A, B,
and E as parameters, the data are fitted to the formula for both the L-waves
and T-waves. The values of the obtained activation energy are

E_L(ultrasonic attenuation, L-wave) = 4.08 eV,   (60)

E_T (ultrasonic attenuation, T-wave) = 2.25 eV.   (61)

The value of E is higher for the L-waves than for the T-waves. The same tendency has also been shown by other authors [83, 84].

In BS experiment, as the light source an Ar ion laser with a longitudinal single mode operated at 514.5 nm wavelength with a power of about 100 mW is used. The Brillouin spectra are obtained in a common 90-degree scattering geometry using a 3-3 pass tandem Fabry-Perot interferometer (Sandercock type). For each spectrum, 1000 channels are assigned and the scanning time is 10 s. The scattered light is detected using a photomultiplier (Hamamatsu R943-02). A photon counter (Princeton Model 1120-1109) is used to count the photopulses, and the signals are accumulated in a personal computer (PC 9801DA). Details of the experimental setup have been presented in [85]. The experiment is performed for PS specimen $1\times1\times1$ mm in size from the room temperature up to a high temperature above T_g, and the heating rate is as low as $3°C/h$.

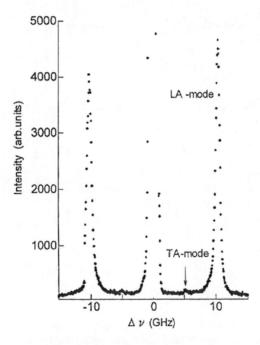

Figure 50. Brillouin spectrum at room temperature showing very strong Rayleigh line, strong LA-mode lines, and weak TA-mode lines.

A typical example of the Brillouin spectrum observed at room temperature is shown in figure 50. The strongest peak of the Rayleigh line appears at the center of the spectrum. Two strong peaks appear on both sides of the Rayleigh line at about $\Delta v = \pm 10$ GHz, and these are the anti-Stokes and Stokes lines representing the longitudinal acoustic phonon mode (LA-mode), respectively. Two very weak lines can also be seen at about $\Delta v = \pm 5$ GHz, and these are the transverse acoustic phonon mode (TA-mode). The temperature of the specimen is increased and the BS experiment is successively carried out. Examples of the observed LA-mode lines at different temperatures are shown in figure 51. The following characteristics of the data can be seen. (a) The position of the scattered line shifts as the temperature is changed. (b) The shape of the line broadens with increasing temperature. (c) The intensity of the line decreases with increasing temperature.

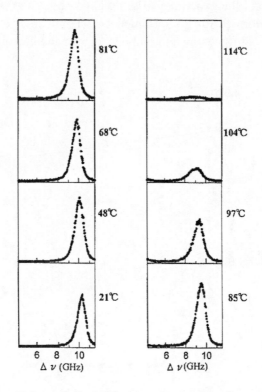

Figure 51. Change of LA-mode line with temperature, where line position decreases, line width increases, and line intensity decreases with temperature.

Temperature dependences of the Brillouin line shifts Δv for both the LA-mode and TA-mode are shown in figure 52. The experimental data for the latter mode are limited because the line is very weak. The inset shows the enlarged figure for the LA-mode near T_g. The line shift is proportional to the velocity of phonons existing in the material, and the phonon velocity is regarded as the velocity of sound when the phonon dispersion can be ignored. In an isotropic medium, the velocities of the longitudinal (L) and transverse (T) waves are given by Eq. 56. The velocity of L-waves is larger than that of T-waves. The monotonic decrease of the sound velocity with increasing temperature arises from the decrease of the elastic modulus due to the lattice anharmonicity. The more rapid decrease of Δv near and above T_g can be an effect of a kind of relaxation.

Using the data as shown in figure 51, the line width W (half-width of the line at the half-maximum) is determined as a function of temperature T. The result is shown in figure 53 for the LA-mode. As the temperature is raised, the value of W changes little at lower temperatures, starts to increase rapidly near T_g, and continues to increase up to temperatures above T_g. The notable temperature dependence of W at higher temperatures is considered to arise from a kind of relaxation. Now, the relation between the line width W and IF is given as [86]:

$$Q^{-1} = 2W/\omega. \qquad\qquad (62)$$

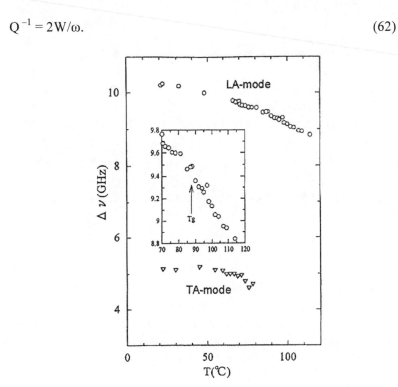

Figure 52. Changes of Brillouin shift Δv for LA- and TA-mode with temperature.

Using Eqs. (57), (58), and (62) the temperature dependence of W is shown to be

$$W \propto \exp(-E/k_B T). \tag{63}$$

From the data in figure 53 the activation energy is determined:

$$E_W \text{ (Brillouin scattering, line width)} = 1.56 \text{ eV}. \tag{64}$$

Meanwhile, from the data as shown in figure 51, the integrated intensity I of the Brillouin line is determined as a function of temperature. In figure 53, the values of the inverse of the intensity, $1/I$, are plotted against temperature T. It can be seen that $1/I$ increases rapidly around T_g, as in the case of W-vs-T. Now we note that

$$I \propto l, \tag{65}$$

where l is the phonon mean-free path. Further, the relation [86]

$$Q^{-1} = V/l\omega \tag{66}$$

is noted, where V is the phonon velocity. Using Eqs. (57), (58), (65), and (66),

$$1/I \propto \exp(-E/k_B T) \tag{67}$$

is finally obtained. Using the experimental data, the activation energy is

$$E_I \text{ (Brillouin scattering, line intensity)} = 1.60 \text{ eV}. \tag{68}$$

It is interesting to see that the two independently determined values, E_W and E_I, are very close to each other. Meanwhile, these values are somewhat smaller than E_L and E_T determined by the US experiment.

In order to reconsider the present results, our previous studies on the viscosity relaxation near T_g for PS are cited. First, the shear viscosity (SV) case is noted [44, 45]. The activation energy obtained in this case is

$$E_{SV} \text{ (shear viscosity)} = 4.80 \text{ eV}. \tag{69}$$

Here, we consider what value of the activation energy is generally observed in relaxation experiment. In glassy materials, the activation energy required for an element (an atom or a group of atoms) to jump is determined by the "potential energy landscape", where a number of potential minima and maxima with different depths and heights are randomly distributed [87]. In the SV case, the relaxing element moves a rather long distance (displacement > 10^{-4} cm) [44], and the element has a chance to encounter a higher potential maximum, then the observed activation energy is high. Next, we consider our IF experiment [51, 52]. The activation energy obtained is

$$E_{IF} \text{ (internal friction)} = 1.57\text{-}0.48 \text{ eV for } f = 0.001\text{-}1.0 \text{ Hz.} \qquad (70)$$

The frequency dependence of activation energy apparently observed can also be explained by the idea of the potential energy landscape [52]. In this case, the relaxation element is considered to be a part of the carbon atom chain in the material. The frequency is lower, the wavelength of the wavy motion of the chain is longer, the displacement of the chain is larger, and the observed activation energy can be higher.

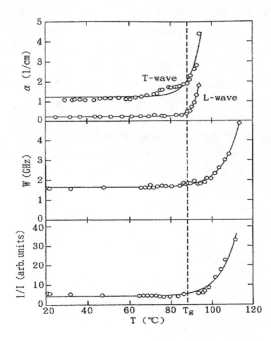

Figure 53. Temperature dependences of sound attenuation, Brillouin line width, and inverse line intensity.

Now the case of the US experiments is considered. The relaxing element can be a single atom. In the case of L-waves, the value of E_L is close to E_{SV}. This means that the displacement amplitude of the atoms is rather large. Meanwhile, in the case of T-waves, the value of E_T is much smaller than E_L. The excitation of strong T-waves is technically difficult, the displacement of atoms is smaller, and the observed activation energy is lower than that in the L-wave case. Next, the case of the BS experiment is considered. Using Eqs. (62) and (66), we can obtain an expression for the phonon mean-free path:

$$l = V/2W, \tag{71}$$

where V is the phonon velocity and W is the line width. Using the observed values, $V \sim 2 \times 10^5$ cm/s and $W \sim 3$ GHz, we find that $l \sim 3 \times 10^{-5}$ cm. This result means that the effective atomic displacement in the BS experiment is rather small. Thus, E_W and E_I are smaller than E_{SV} or E_L.

It is shown that the activation energy is mainly determined by the potential energy landscape in the material, and may not primarily be dependent on the frequency of the probe. This is a static aspect of the relaxation. However, the effect of the probe frequency can be seen as another aspect of the relaxation. See again the data shown in figure 53. The rapid exponential increase representing the relaxation occurs in a higher temperature range for higher frequency probe. This is also consistent with the results of our previous studies. In the SV measurement the Arrhenius behavior starts at a temperature close to T_g. In the low-frequency IF measurement the rapid exponential increase of IF seems to start at a temperature somewhat higher than T_g. Thus, the hopping region of the viscoelastic relaxation is shifted to higher temperature range when the high-frequency probe is used. This is a dynamical aspect of the relaxation observed experimentally.

REFERENCES

[1]   Y. Hiki: *J. Phys. Soc. Jpn.* 13 (1958) 1138.
[2]   Y. Hiki: *J. Phys. Soc. Jpn.* 14 (1959) 590.
[3]   J. Marx: *Rev. Sci. Instr.* 22 (1951) 503.
[4]   A. Granato and K. Lücke: *J. Appl. Phys.* 27 (1956) 583, 789.
[5]   A. H. Cottrell and B. A. Bilby: *Proc. Phys. Soc. London A* 62 (1949) 49.
[6]   Y. Hiki: *J. Phys. Soc. Jpn.* 15 (1960) 586.
[7]   Y. Hiki: *J. Phys. Soc. Jpn.* 16 (1961) 664.
[8]   T. Kosugi, Y. Kogure and Y. Hiki: *J. Phys. Soc. Jpn.* 54 (1985) 2565.
[9]   T. Kosugi, Y. Kogure, Y. Hiki and T. Kino: *J. Phys. Soc. Jpn.* 55 (1986) 1203.
[10] Y. Kogure, T. Kosugi and Y. Hiki: *J. Phys. Soc. Jpn.* 54 (1985) 4592.
[11] G. Leibfried: *Z. Phys.* 127 (1950) 344.
[12] W. P. Mason: *J. Acoust. Soc. Am.* 32 (1960) 458.
[13] T. Ninomiya: *J. Phys. Soc. Jpn.* 25 (1968) 830.
[14] F. Tsuruoka and Y. Hiki: *Phys. Rev. B* 20 (1979) 2702.
[15] Y. Hiki and F. Tsuruoka: Phys. Rev. B 27 (1983) 696.
[16] A. Seeger: *Report on the Conference of Defects in Crystalline Solids, Bristol, 1954* (Physical Society, London, 1955) p. 391.
[17] Y. Hiki, F. Tsuruoka and Y. Kogure: *J. Phys. Soc. Jpn.* 57 (1988) 134.
[18] H. Kamioka: *J. Phys. Soc. Jpn.* 52 (1983) 2433.
[19] H. Kamioka: *J. Phys. Soc. Jpn.* 53 (1984) 1349.
[20] F. R. N. Nabarro: *Theory of Crystal Dislocations* (Clarendon, Oxford, 1967).
[21] J. Tamura, Y. Kogure and Y. Hiki: *J. Phys. Soc. Jpn.* 55 (1986) 3445.
[22] Y. Kogure, H. Endo and Y. Hiki: *J. Phys. Soc. Jpn.* 56 (1987) 1404.
[23] A. B. Pippard: *Philos. Mag.* 46 (1955) 1104.

[24] A. B. Pippard: *Proc. R. Soc. London A* 257 (1960) 165.

[25] I. Nakamichi and T. Kino: *J. Phys. Soc. Jpn.* 49 (1980) 1350.

[26] J. Bardeen, L. N. Cooper and J. R. Schrieffer: *Phys. Rev.* 108 (1957) 1175.

[27] Y. Kogure, N. Takeuchi, Y. Hiki, K. Mizuno and T. Kino: *J. Phys. Soc. Jpn.* 54 (1985) 3506.

[28] Y. Hiki and A. V. Granato: *Phys. Rev.* 144 (1966) 411.

[29] R. N. Thurston and K. Brugger: *Phys. Rev.* 133 (1964) A1604.

[30] K. Brugger: *J. Appl. Phys.* 36 (1965) 768.

[31] Y. Hiki, J. F. Thomas, Jr. and A. V. Granato: *Phys. Rev.* 153 (1967) 764.

[32] H. Soma and Y. Hiki: *J. Phys. Soc. Jpn.* 37 (1974) 544.

[33] S. Mori and Y. Hiki: *J. Phys. Soc. Jpn.* 45 (1978) 1449.

[34] T. Maruyama and Y. Hiki: *J. Phys. Soc. Jpn.* 35 (1973) 1142.

[35] Y. Hiki and K. Mukai: *J. Phys. Soc. Jpn.* 34 (1973) 454.

[36] H. Kobayashi and Y. Hiki: *Phys. Rev. B* 7 (1973) 594.

[37] H. Kobayashi and Y. Hiki: *Jpn. J. Appl. Phys.* 11 (1972) 738.

[38] Y. Hiki, H. Takahashi and H. Kobayashi: *Solid State Ionics* 53-56 (1992) 1157.

[39] Y. Hiki, H. Kobayashi and H. Takahashi: *J. Phys. IV* 6 (1996) C8-609.

[40] H. Kobayashi, Y. Hiki and H. Takahashi: *J. Appl. Phys.* 80 (1996) 122.

[41] Y. Hiki, H. Kobayashi, H. Takahashi and Y. Kogure: *Prog. Theor. Phys. Suppl.* No. 126 (1997) 245.

[42] H. Takahashi, Y. Hiki and H. Kobayashi: *J. Appl. Phys.* 84 (1998) 213.

[43] Y. Hiki, H. Kobayashi and H. Takahashi: *J. Alloys Compd.* 310 (2000) 378.

[44] H. Kobayashi, H. Takahashi and Y. Hiki: *J. Non-Cryst. Solids* 290 (2001) 32.

[45] Y. Hiki and Y. Kogure: *J. Non-Cryst. Solids* 315 (2003) 63.

[46] C.Zener: *Elasticity and Anelasticity of Metals* (Univ. of Chicago, Chicago, 1948).

[47] A. S. Nowick and B. S. Berry: *Anelastic Relaxation in Crystalline Solids* (Academic, New York, 1972).

[48] G. R.Strobl: *The Physics of Polymers* (Springer, Berlin, 1997).

[49] I. Gutzow and J. Schmelzer: *The Vitreous State* (Springer, Berlin, 1995).

[50] C. A. Angell: *J. Phys. Chem. Solids* 49 (1988) 863.

[51] Y. Hiki, Y. Maeda, A. Maesono and T. Kosugi: *J. Non-Cryst. Solids* 312-314 (2002) 613.

[52] Y. Hiki and T. Kosugi: *J. Non-Cryst. Solids* 351 (2005) 1300.

[53] G. Fantozzi: *Mechanical Spectroscopy* (Trans Tech Publications, Zuerich, 2001).

[54] P. A. Tanguy, L. Choplin and P. Hurez: *Polym. Eng. Sci.* 28 (1988) 529.

[55] Y. Hiki and Y. Kogure: Recent Res. Devel. *Non-Cryst. Solids* 3 (2003) 199.

[56] Y. Kogure and Y. Hiki: *J. Appl. Phys.* 88 (2000) 582.

[57] W. L. Johnson: *Mater. Sci. Forum* 225/227 (1996) 35.

[58] W. L. Johnson: *MRS Bull.October* 21 (1996) 42.

[59] A. Inoue: *Mater. Trans. JIM* 36 (1995) 866.

[60] A. Inoue: *Acta Mater.* 48 (2000) 279.

[61] Y. Hiki, T. Aida and S. Takeuchi: *J. Phys. Soc. Jpn.* 76 (2007) 114601.

[62] Y. Hiki, T. Yagi, T. Aida and S. Takeuchi: *J. Alloys Compd.* 355 (2003) 42.

[63] Y. Hiki, T. Aida and S. Takeuchi: *Proc. 3rd Int. Sym. on Slow Dynamics in Complex Systems* (American Institute of Physics, New York, 2004) p. 661.

[64] Y. Hiki, T. Yagi, T. Aida and S. Takeuchi: *Mat. Sci. Eng. A* 370 (2004) 302.

[65] S. Tyagi and A.E. Load, Jr.: *J. Non-Cryst. Solids* 30 (1979) 273.

[66] T. Soshiroda, M. Koiwa and T. Masumoto: *J. Non-Cryst. Solids* 22 (1976) 173.

[67] N. Morito: *Mat. Sci. Eng.* 60 (1983) 261.

[68] V. A. Khonik: *J. Physique IV* 6 (1996) C8-591.

[69] H. S. Chen, H. J. Leamy and M. Barmatz: *J. Non-Cryst. Solids* 5 (1971) 444.

[70] Y. Hiki, M. Tanahashi and S. Takeuchi: *Mat. Sci. Eng. A* 442 (2006) 287.

[71] Y. Hiki, M. Tanahashi and S. Takeuchi: *J. Non-Cryst. Solids* 354 (2008) 1780.

[72] Y. Hiki, M. Tanahashi, R. Tamura, S. Takeuchi and H. Takahashi: *J. Phys. Condens. Matter* 19 (2007) 205147.

[73] T. A. Waniuk, R. Busch, A. Masuhr and W. L. Johnson: *Acta mater.* 46 (1998) 5229.

[74] Y. Hiki, M. Tanahashi and S. Takeuchi: *J. Non-Cryst. Solids* 354 (2008) 994.

[75] P. Wen, D. Q. Zhao, M. X. Pan, W. H. Wang, J. P. Shui and Y. P. Sun: *Intermetallics* 12 (2004) 1245.

[76] T. Yagi, T. Imai, R. Tamura and S. Takeuchi: *Mat. Sci. Eng. A* 370 (2004) 264.

[77] U. Geyer, S. Schneider, W. L. Johnson, Y. Qiu, T. A. Tombrello and M.-P. Macht: *Phys. Rev. Lett.* 75 (1995) 2364.

[78] U. Geyer, W. L. Johnson, S. Schneider, Y. Qui, T. A. Tombrello and M.-P. Macht: *Appl. Phys. Lett.* 69 (1996) 2492.

[79] Y. Takagi, T. Hosokawa, K. Hoshikawa, H. Kobayasdhi and Y. Hiki: *J. A. Phys. Soc. Jpn.* 76 (2007) 024604.

[80] H. J. McSkimin: *Physical Acoustics* Vol. I-Part A, edited by W. P. Mason (Academic, New York, 1964).

[81] R. Truell, C. Elbaum and B. B. Chick: *Ultrasonic Methods in Solid State Physics* (Academic, New York, 1969).

[82] L. D. Landau and E. M. Lifshitz: *Theory of Elasticity* (English Edition, Pergamon, Oxford, 1959).

[83] R. Kono: *J. Phys. Soc. Jpn.* 15 (1960) 718.

[84] R. S. Marvin and J. E. McKinney: *Physical Acoustics* Vol. II-Part B, edited by W. P. Mason (Academic, New York, 1965).

[85] A. Nagase, Y. Takeuchi and Y. Takagi: *Jpn. J. Appl. Phys.* 35 (1996) 2903.

[86] U. Buchenau: *Phys. Rev. B* 63 (2001) 104203.

[87] C. A. Angell: *J. Non-Cryst. Solids* 131-133 (1991) 13.

INDEX